Advanced Methods for Conducting

Online Behavioral Research

Advanced Methods
for Conducting

Online
Behavioral
Research

Edited by

Samuel D. Gosling
John A. Johnson

American Psychological Association • Washington, DC

Second Printing, March 2011

Published by
American Psychological Association
750 First Street, NE
Washington, DC 20002
www.apa.org

To order
APA Order Department
P.O. Box 92984
Washington, DC 20090-2984
Tel: (800) 374-2721; Direct: (202) 336-5510
Fax: (202) 336-5502; TDD/TTY: (202) 336-6123
Online: www.apa.org/books/
E-mail: order@apa.org

In the U.K., Europe, Africa, and the Middle East, copies may be ordered from
American Psychological Association
3 Henrietta Street
Covent Garden, London
WC2E 8LU England

Typeset in Meridien by Circle Graphics, Inc., Columbia, MD

Printer: Edwards Brothers, Ann Arbor, MI
Cover Designer: Mercury Publishing Services, Rockville, MD

The opinions and statements published are the responsibility of the authors, and such opinions and statements do not necessarily represent the policies of the American Psychological Association.

Library of Congress Cataloging-in-Publication Data

Advanced methods for conducting online behavioral research / edited by Samuel D. Gosling and John A. Johnson. — 1st ed.
   p. cm.
 Includes bibliographical references.
 ISBN-13: 978-1-4338-0695-7
 ISBN-10: 1-4338-0695-9
 1. Psychology—Research—Data processing. 2. Internet research. 3. Psychology—Research—Methodology. I. Gosling, Sam. II. Johnson, John A. (John Anthony), 1953-

 BF76.6.I57A38 2010
 150.285'4678—dc22

                                    2009024842

British Library Cataloguing-in-Publication Data
A CIP record is available from the British Library.

Printed in the United States of America
First Edition

Contents

v

Contributors

Michael H. Birnbaum, Decision Research Center, California State University, Fullerton

Tom Buchanan, University of Westminster, London, England

Alastair J. Gill, Northwestern University, Evanston, IL

Anja S. Göritz, University of Würzburg, Germany

Samuel D. Gosling, University of Texas at Austin

Lars Kaczmirek, GESIS—Leibniz Institute for the Social Sciences, Mannheim, Germany

John H. Krantz, Hanover College, Hanover, IN

John A. Johnson, Pennsylvania State University, Dubois

Elizabeth Mazur, Pennsylvania State University, Greater Allegheny

Matthias R. Mehl, University of Arizona, Tucson

Wolfgang Neubarth, GESIS—Leibniz Institute for the Social Sciences, Mannheim, Germany (formerly ZUMA)

Ulf-Dietrich Reips, Universidad de Deusto, (d) IKERBASQUE, Basque Foundation for Science, Basque, Spain

Stefan Schipolowski, Humboldt University of Berlin, Germany

Ulrich Schroeders, Humboldt University of Berlin, Germany

Olaf Thiele, University of Mannheim, Germany

Tracy L. Tuten, East Carolina University, Greenville, NC

Sonja Utz, VU University Amsterdam, The Netherlands

Simine Vazire, Washington University in St. Louis

Oliver Wilhelm, Humboldt University of Berlin, Germany

John E. Williams, University of Northern Iowa, Cedar Falls

GETTING STARTED

John A. Johnson and Samuel D. Gosling

How to Use This Book 1

V ery simply, the aim of this book is to help you get research done. If you have already made the decision to use the Internet as part of your research, and if you have a rudimentary understanding of Web-based research methods, then this book is for you. Our goal is to provide you with a practical step-by-step guide—something that can sit by you on the desk as you develop and implement your study. The chapters, which cover a broad array of topics encountered by social science researchers, are short and simple. The authors are world experts on their topics who draw on their extensive experience (read: mistakes!) to offer advice on how to avoid common pitfalls and implement your study as efficiently as possible.

Before You Use This Book

A number of chapters in this book assume at least a basic knowledge of Web-page construction. If you are unfamiliar with hypertext markup language (HTML), Common Gateway Interface (CGI) scripts, and the process of uploading Web pages and maintaining a Web site, we strongly recommend that you first read Fraley's (2004) outstanding book *How to Conduct Behavioral Research Over the Internet: A Beginner's Guide to HTML and CGI/Perl.* Our book is aimed one level above Fraley's book, at researchers who want more information on more specific topics and on the numerous new methods (e.g., text analysis) that are constantly being developed to take advantage of emerging research opportunities. The goal of this volume is to present the varied methods in short, practically oriented chapters that will allow investigators to integrate the methods into their own behavioral research.

Overview of Book Structure

This book is organized into five parts. Part I, Getting Started, contains the introductory chapter that you are reading now and a chapter by Michael H. Birnbaum that summarizes general issues to consider in Internet-based research. Part II, Considerations When Designing Web Pages, contains three chapters on the construction of Web pages. The chapter by Ulf-Dietrich Reips describes the optimal design of Web pages. The next chapter, by John H. Krantz and John E. Williams, describes how to incorporate graphics, photographs, and dynamic media into Web pages. The final chapter in this part, by Wolfgang Neubarth, explains how to incorporate moveable, drag-and-drop objects into Web pages to allow participants to make visual rankings and ratings.

Part III, Studying Internet Behavior, considers methods of studying human activities that occur on the Internet. Researchers are just beginning to examine the degree to which phenomena such as self-expression and formation of coalitions follow the same patterns as those seen in the real world or whether new psychological principles are needed to explain behavior in virtual worlds. The chapter by Elizabeth Mazur presents methods for analyzing social networking Web sites and blogs. Sonja Utz's chapter brings the researcher a little closer to these new Internet behavioral phenomena as a participant observer. Finally, Matthias R. Mehl and Alastair J. Gill describe a method for automatic analysis of text that is posted on the World Wide Web.

Part IV, Transporting Traditional Methodologies to the Web, contains five chapters, each describing how a long-standing research method can be used on the Web. These methods include ability testing (Ulrich Schroeders and Oliver Wilhelm), personality self-reports (John A. Johnson), informant reports (Simine Vazire), surveys (Tracy L. Tuten), and experiments (Ulf-Dietrich Reips and John H. Krantz).

Part V, Cross-Cutting Issues, covers concerns that are common to all researchers, regardless of their specific research topics or methods.

In this part are chapters on using incentives to recruit and retain research participants (Anja S. Göritz), securing and protecting one's data (Olaf Thiele and Lars Kaczmirek), and conducting research in an ethical manner (Tom Buchanan and John E. Williams).

Structure of Each Chapter

Each chapter in this volume is designed to be a self-contained tutorial on a particular topic in Internet-based research. The chapters may therefore be read in any order. Readers can simply scan the table of contents for topics of greatest relevance to their own research and turn directly to the chapters dealing with those topics.

We have asked all contributors to introduce their chapters in such a way that a reader will be able to determine after reading one or two pages whether a chapter will be relevant to his or her research. Following a brief introduction, each chapter then gets down to the specifics of implementing a method for addressing an issue. Most of the information that is needed to implement a method is contained in the chapter itself. Many of the chapters also provide up-to-date links, additional details, examples, and samples of computer code that can be copied and modified to the researcher's needs; these items are provided at the supplementary materials Web site at http://www.apa.org/books/resources/gosling. Each chapter also contains a list of additional resources for researchers who wish to explore certain topics in even greater depth.

Pros and Cons of Internet-Based Versus Traditional Research Methods

Initial attempts at transporting traditional research methods to the Internet were greeted with a healthy dose of skepticism. Quite reasonably, journal editors and reviewers had a number of concerns about method artifacts and sampling issues. Gosling, Vazire, Srivastava, and John (2004) empirically addressed six of the most common concerns by comparing a large Internet sample with a year's worth of conventional samples used in one year's worth of studies published in the *Journal of Social and Personality Psychology,* the field's top outlet. Their analyses suggested that, compared with conventional samples, Internet samples are relatively diverse with respect to gender, socioeconomic status, geographic region, and age.

Moreover, Internet findings generalize across presentation formats, are not adversely affected by nonserious or repeat responders, and are consistent with findings from traditional methods. Similar conclusions were reached by other reviews addressing the validity of Internet research (e.g., Krantz & Dalal, 2000).

Before you start your research, we think that it is important to keep in perspective the advantages and disadvantages of Internet-based versus traditional research methods. A major advantage of computer-assisted data collection is the efficiency and accuracy with which traditional forms of data could be collected (e.g., surveys, informant reports, reaction time experiments). Prior to the widespread availability of computers, all psychological data had to be collected and recorded by hand, opening the possibility of clerical errors. The advent of scan sheets and recording equipment connected directly to computers may have reduced some recording errors. But the use of the Web allows for the added advantage of data collection from around the world without the delays and expense of land-based mail. The Internet has eliminated the time and space constraints of traditional data collection. Furthermore, the validity of protocols can be checked instantly; the data can be stored automatically; and feedback, which serves as a major incentive for participation, can be delivered instantaneously to research participants.

Another advantage of Internet-based research is the potential for obtaining very large, diverse samples from around the world. By providing personalized automated feedback, researchers in personality (e.g., John Johnson, Sam Gosling, Jeff Potter) and experimental social psychology (e.g., Brian Nosek) have been able to collect data from hundreds of thousands of participants, samples previously unheard of in psychological research. Also, by providing anonymity or by covert observation techniques, researchers have been able to study groups such as sex offenders or White supremacists who might not participate in studies in real life (e.g., Glaser, Dixit, & Green, 2002). Finally, rich multimedia can be included in ability tests, questionnaires, surveys, and experiments, creating a more life-like and ecologically valid environment than ordinarily found in pencil-and-paper measures.

Internet-based research has many pros, but one must also consider the cons. The central problem stemming from the physical disconnect between researcher and participant is lack of control over the assessment or experimental setting. Because researchers are not physically present, they cannot easily assess the alertness and attentiveness of the participants. Furthermore, they will have difficulty in immediately answering questions from participants about the procedure. Because they are not directly observing the research participants, they cannot be aware of possible distractions to the participants such as eating, drinking, television, music, conversations with friends, and the perusal of other Web sites.

Internet users, especially young Internet users, are notorious for multi-tasking when they are logged on, and this could have adverse effects on the quality of Internet-based data. In the case of ability testing, with all of the information on the Internet at their disposal, what is to keep participants from cheating?

Although the research by Gosling et al. (2004) has allayed many fears about the quality of Internet versus real-life samples, participation in Internet-based research is obviously restricted to people who have access to the Internet; who know how to use a Web browser; and in some kinds of research, who have a functioning e-mail address or instant messaging capability. Excluded from Internet research will be people who are computer phobic, who cannot afford a computer and Internet service and have no public access to the Internet, and who are uninterested in learning how to browse the Web.

Finally, learning to construct Web pages, write program scripts, manage computer data bases, and engage in all of the other activities involved in starting up online research is admittedly time consuming. Entire new sets of skills must be acquired, practiced, and polished. We hope that this book will help you to acquire these skills so that you may enjoy the many advantages of Internet-based research.

References

Fraley, R. C. (2004). *How to conduct behavioral research over the Internet: A beginner's guide to HTML and CGI/Perl.* New York: Guilford Press.

Glaser, J., Dixit, J., & Green, D. P. (2002). Studying hate crime with the Internet: What makes racists advocate racial violence? *Journal of Social Issues, 58,* 177–193.

Gosling, S. D., Vazire, S., Srivastava, S., & John, O. P. (2004). Should we trust Web-based studies? A comparative analysis of six preconceptions about Internet questionnaires. *American Psychologist, 59,* 93–104.

Krantz, J. H., & Dalal, R. (2000). Validity of Web-based psychological research. In M. H. Birnbaum (Ed.), *Psychological experiments on the Internet* (pp. 35–60). San Diego, CA: Academic Press.

Michael H. Birnbaum

An Overview of Major Techniques of Web-Based Research

2

This chapter presents a summary of major methods useful to Web-based research. It is intended to help researchers decide what techniques they need to learn for their particular type of research.

HyperText Markup Language

The most basic method for formatting, organizing, and linking information on the Web is *hypertext markup language* (HTML). This language contains instructions (commands) called *tags* that tell the client's (i.e., your research participant's) browser how to display information contained in files stored on one or more server(s) on the Web. If you are planning to do research via the World Wide Web (WWW), you need to learn basic HTML. Many free tutorials on the Web teach HTML; there are also books that you can use to teach yourself HTML (Birnbaum, 2001; Fraley, 2004).

HTML files are simple text files that can be created in a simple text editor such as Notepad. HTML files can also be created and edited by Web page development programs like Dreamweaver, a commercial program available from Adobe (originally created by Macromedia). I advise my students to learn HTML before they even decide whether to purchase such commercial programs. Knowledge of how HTML works helps you to understand what the development programs are doing; it helps you avoid terrible errors that can result in data loss and ruined studies that can occur when an ignorant person uses a program he or she does not understand; and it helps you decide what software, if any, you need. When you understand HTML yourself, you will find it is often easier to work directly in HTML rather than struggle with a "what you see is what you get" type of editor. The problem is that what you don't see may result in a failure to get your data in a usable form, if at all.

HTML allows one to do any study that could have been done with paper and pencil, a slide projector, video recorder, and more. One can insert not only graphics and photos in a survey or experiment (which could have also been done through a paper questionnaire) but also sound and video (which are not easy to implement by paper). For a discussion of media that can be delivered via the WWW, see Krantz (2001). Instructional materials by Krantz for the creation and editing of media are available at http://psych.hanover.edu/NSFATI/. Krantz was one of the first people to conduct research via the Web (for a discussion of validity of Web studies, see Krantz & Dalal, 2002,), and he maintains a site where experiments can be listed as a way to recruit volunteer participants in Web-based research: http://psych.hanover.edu/Research/exponnet.html.

Basic HTML permits use of one question that will either include or skip other items. For example, the question "Do you smoke?" might be used to link to a series of questions about smoking or to skip them, depending on how the participant responds. One can also use HTML to have a person click on his or her birth month to create assignment of participants to conditions. Skipping items and assigning participants to between-subjects conditions are examples of what can be done with hyperlink tags.

Web Forms

A very powerful technique that is built into HTML is known as *Web forms*. For example, this technique allows the client to respond to items in a questionnaire by typing in answers to questions, entering numbers, clicking along rating scales, or choosing from pull-down lists. This method can be used to send data to be saved in a file on a server, from which it can be later downloaded and imported to a statistical package, such as SPSS, or opened in a spreadsheet program, such as Excel. The technique can also be used to append data to the log file of the server or to send data via e-mail to an e-mail account.

Exhibit 2.1 shows a basic Web page containing a form that sends two data to a *Common Gateway Interface* (CGI; a standard for computers to exchange information via the Internet) script. The first datum is the hidden variable whose value is "MyTestAge," which identifies the study. The second datum is the person's age (or whatever he or she typed in the box). In this case, the script is written in the programming language Perl (Practical Extraction and Reporting Language). You can test this example on the supplementary Web site for this chapter.

The <form> tag has an <action> specified. The action in Exhibit 2.1 sends the data to the address of a CGI file written in Perl that saves the data to a Web site, which can be viewed at the supplementary Web site for this chapter.

Your data will be the last information contained in the data file on the server that was written by the CGI. This data file is available to be read on the Web. Normally, one would not allow permission for a data file to be read; however, in this case, the file has been given no protection, so that you can test the example. The script also redirects the user to a "thank you" page.

To send data to the server's log file instead of to this data file, you could change the <form> tag as follows:

<form method=get action=thanks.htm>

EXHIBIT 2.1

A bare bones Web form. This Web form requests just one datum from the participant: age. When the "submit" button is clicked, this form sends its data to a CGI script called "simple.pl" in the folder called "cgi-bin" in the Web site (see Exhibit 2.2). That script saves the data to the server.

```
<html>
<head>
<title>My First Form</title>
</head>
<body>
Please answer this question.
<form method= "post" action= "http://ati-birnbaum-2008.netfirms.com/
cgi-bin/simple.pl">
<input type= "hidden" name= "00exp" value= "MyTestAge">
<p> What is your age?
<input type= "text" name= "01age" size= "2" maxlength= "3">
<p><input type= "submit" value= "Send the Data">
</form>
</body>
</html>
```

This method (*get*) will append the data to the next URL and to the server's log file, and in this case, it also redirects the participant to a "thank you" page in the same folder as the survey.

CGI Scripts

Exhibit 2.2 presents a simplified version of a Perl script that was originally written by Schmidt to emulate GCI scripts that were created by PolyForm, a program that is no longer supported. This CGI program organizes the data in the order of leading digits on the variable names (e.g., as in Exhibit 2.1). That aspect of the script means that a set of questions can be placed in many different random orders, but the data will return in a fixed order for analysis. That trick was used by Birnbaum (2000b) in his FactorWiz program, which creates the HTML for studies with randomized orders of factorial combinations.

EXHIBIT 2.2

A simple Perl script that saves data (simple1.pl). This file is placed in the Web site folder called "cgi-bin." Data are saved into a file named "data.txt," which is inside a folder named "data." This script sorts data according to leading digits that precede the input variable names. Each datum appears in quotes, separated by commas. This type of file can be easily imported to spreadsheet programs such as Excel. The word *end* is placed at the end of each data record. The user is redirected to a thank-you message in the file thanks.htm.

```
#!C:/perl/bin/perl.exe
$path_to_datafile = ". ./www/data";
$redirect_to = ". ./thanks.htm";
use CGI;
$query = new CGI;
#open data file (in folder data, data.txt) and save data
open(INFO, ">>$path_to_datafile/data.txt");
foreach $key (sort($query->param))
{
$value = $query->param($key);
print INFO "\ "$value\",";
}
print INFO "\ "end\ "\n";
close (INFO);
print $query->redirect($redirect_to);
```

To learn how to install a generic Perl script to save data to your own server, read the instructions on the supplementary site for this chapter. This Perl script can be used with all of the examples in Birnbaum (2001).

Another CGI programming language is PHP (originally standing for Personal Home Page, now standing for PHP: Hypertext Preprocessor), a set of powerful hypertext processing tools used to create dynamic Web pages that can respond to and interact with the client. PHP is available from http://php.net/. A generic form processor in PHP is provided in Göritz and Birnbaum (2005), which also works with the examples in Birnbaum (2001).

Organization of a Web Site

Table 2.1 shows the organization of the bare bones Web site that has been created at the supplementary site for this chapter. One of the two main folders is called "www" and contains the HTML files that can be viewed via the Web. Depending on the server and its configuration, this folder might be called "htdocs" or "website," for example, instead of "www." It contains a home page called "index.htm." When a person enters the URL for this site in their browser, the file index.htm in this folder will be displayed.

TABLE 2.1

Organization of a Simple Web Site

www (folder)	cgi-bin (folder)
index.htm	simple.pl (see Exhibit 2.2)
Listing1.htm (Exhibit 2.1)	generic.pl
Survey_1.htm	
thanks.htm	
data (folder)	
data.txt (file inside the data folder)	

Note. This Web site has two main folders: www (which contains the HTML documents that can be viewed via the Web) and cgi-bin (which contains the CGI scripts that save data to the server). The file index.htm is the home page that will be displayed when a person types in the URL of the Web site. This page might contain text describing the study, inviting people to participate, and a link to the study. Two studies, survey1.htm and survey2.htm are listed. The file thanks.htm contains a "thank you" and debriefing message. It could also contain links to other pages (e.g., other studies to do, a report describing the research program). The data folder contains the data.txt file, to which the data are appended. There could be other data files as well. The data folder and its files can be password protected.

The www folder also contains two surveys, "Listing1.htm" and "survey_1.htm." These files are linked via hyperlinks from the home page, which would normally contain information about the studies and an invitation to participate. This folder also contains the file thanks.htm, to which the participant is directed when he or she completes the survey and submits the data. The thank you page might contain links to other surveys, to debriefing materials, or to other resources.

The other main folder is the cgi-bin folder, which in this case contains two scripts written in Perl. The first of these is "simple.pl," which contains Exhibit 2.2. Also included in this folder is the generic script, "generic.pl," that allows one to record the date and time of the submission and to gather the remote Internet Protocol (IP) address of the computer used by the participant, as well as other information. These scripts are saved as simple text files with the extension, ".pl" (i.e., ".PL"), which indicates a Perl program.

A researcher might recruit participants by various methods to visit the home page. From there, the person would be invited to click a link to the appropriate survey or experiment. When the person completes the study and clicks the "submit" button, the data are sent to the CGI (e.g., the generic Perl script), which saves the data in a file (in this case, it is called "data.txt" in a folder named "data"), and the participant is redirected to the page (in this case, "thanks.htm"), which would contain a "thank you" message and appropriate debriefing, including perhaps links to read more about the research topic. In the bare bones Web site, a link has been provided to the data file, so people testing the system can check that their data arrived correctly. (After scripts have been tested, the link to the data would normally be removed and the permissions on the data folder and data files would be set so that only the experimenter can view them by signing in with a password.)

What Server to Use?

The term *server* refers to both (a) the computer that houses and "serves" (delivers) the HTML files to the participant and (b) the computer software that performs those functions. It is possible that two servers might be used: One server might host the surveys and experiments, and the other might be used to save the data and redirect the participant to the "thank you" message or next part of the study. Usually, however, the same server would be used for both the survey and the programs that save data.

Perhaps your university or department maintains a server that can host your surveys and experiments. You may have people on your campus who can provide helpful and knowledgeable tech support. To upload your files to your campus server from a remote site (such as your home computer, for example), you can use a File Transfer Protocol (FTP) program. This method requires that your department or university provide you an account and an FTP password. Many free programs perform FTP, which can be found by searching on sites such as http://www.download.com for *FTP* or using search engines such as http://www.google.com to find such sites for shareware and freeware. If you are allowed to work on the server directly, you can simply copy your files from a flash drive or other data storage device to your folder of Web files on the server.

To save data from Web surveys and experiments on a server, you create Web pages that collect and send data to a CGI script on the server that organizes and saves the data. To create surveys in HTML, you might use SurveyWiz (Birnbaum, 2000b), which is a very simple-to-learn program that creates the HTML for a survey consisting of text boxes, true–false, multiple choice, and rating scale items. To try out SurveyWiz, visit the supplementary Web site for this chapter. For your CGI script, you can use a Perl script such the one described above, "generic.pl." Alternatively, you might use the PHP generic form processor (Görtiz & Birnbaum, 2005).

Unfortunately, a problem that some researchers encounter at universities is that technicians who are placed in charge of servers may not be knowledgeable about Web research and may also be overly concerned about security or their own personal control of the site. You might not be allowed to work directly on the server, to see its log files, or to install Perl programs or other such tasks that might be needed for your research.

Sometimes technicians are asked to enforce directives from administrators who want to dictate the appearance or even the content of all Web pages hosted by the university. Some universities create rules about what can or cannot be put on "their" Web sites. Other universities require that half or more of the space of every Web page be filled with promotional banners and insignias of the university, to make everything look official and uniform.

To avoid such problems in dealing with universities that do not allow academic freedom on the Internet, one can use a private Web hosting service. Some of these providers offer basic Web hosting for free, which usually means that commercial banner ads will be placed in some, if not all, of the Web pages in a site. For example, in 2009, the company www.netfirms.com offered basic Web hosting free, including the option to include Perl scripts, as well as other features. To find this service, visit http://www.netfirms.com/web-hosting/web-hosting-basic/. My example site, at the supplementary Web site for this chapter, also includes instructions on installing a generic Perl script and illustrates how your site might look (including the advertisements) when hosted this way.

For a fee of about $5 per month, you could have a commercial Web site without the ads and conduct your research there. For many investigators, a commercial site costing $5 per month is well worth the expense.

Such an approach provides one freedom of expression such as is no longer available at many universities. Furthermore, you can relax: Others will make sure that the site is up and running and that it is secure, relieving you of the daily needs to make sure that the power has not gone out, that the system has not crashed, and that hackers are not attacking your site. No one will ask you to change fonts or to install university logos on all pages. You won't need permission from the technical staff every time you want to upload or change your files.

Some people, however, prefer to run their own server. This solution gives you maximal control over all of the files on your computer and is a preferred method if you will be using a database that interacts with your participants or requires a lot of server-side programs. This approach requires that you have a good connection to the Internet (through either your university or a commercial Internet service provider), you have server software, you know how to use these programs, and you can maintain the server and make sure that you keep everything running.

Fortunately, the best server software is free. The Apache HTTP Server is included (installed) with all new Macintosh computers. If you have a Mac, all you have to do is turn on the server that has already been installed and make a few simple adjustments, which are described in Birnbaum and Reips (2005). If you have a PC running Windows or Linux, you can download the free Apache software from http://www.apache.org/.

Göritz (2004) has described how to install a package of free software, including Apache HTTP Server, along with PHP and MySQL, which are used to collect, organize and store data in a database. See Göritz and Birnbaum (2005) for a generic script in PHP that can be used to emulate the scripts needed to work with Birnbaum's (2000a, 2001) examples. Links to Göritz's resources are given in the next section, which deals with the topic of programming the server.

Server-Side Programming

Server-side programs can feed dynamic material to the client on the basis of what the client puts in. The term "server-side" means that these programs run on the server rather than on the participant's computer. The two most popular server-side languages are PHP and Perl. The PHP Hypertext Preprocessor is a server-side programming language that allows Web developers to create dynamic content that interacts with databases. Like Perl, PHP is free but requires some study to learn to use its full powers. Neither you nor your participant need buy anything.

These are called "server-side" programming languages because the programs are executed by the server rather than by the client's computer. That means that these programs will work for all participants, no matter what type of computer they have, as long as they can access the Internet and display Web pages.

Both of these free languages are described in tutorials that can be found on the Web. Books are also available on these languages. Perl is available from http://www.perl.org, and PHP can be found at http://www.php.net. For a tutorial on PHP, see http://us.php.net/tut.php.

Schwarz (1998) presented an introductory tutorial on Perl, and Fraley's (2004) book on Web experimentation contains lessons in Perl. Fraley showed how to use Perl to accomplish common tasks in behavioral research such as random assignment of participants to conditions and random ordering of items within a study. Many of his examples and other resources can be found at http://www.web-research-design.net/.

If you plan to collect surveys and experiments from participants and then simply analyze the results, you do not need any more than a generic script that saves the data to a file on the server (see, e.g., Exhibit 2.2). You do not need to run your own server. Your data can be saved in a simple data file that can be imported to statistical programs like SPSS.

Schmidt (1997) has written software called WWW Survey Assistant, which creates both a survey and a corresponding CGI script in Perl to collect data from surveys and save them to a server. For information on his approach, see http://www.mohsho.com/s_ware/how.html.

If you intend to interact with your research participants in a dynamic manner (or over an extended period of time), you will want to run your own server and use server-side programming to save data in a database. The most popular database software is MySQL (My Structured Query Language), which is also open source and free: http://www.MySQL.org. Open source means that program source code can be seen by everyone. An advantage of open source, free software is that thousands of people will test and evaluate the software and work together to improve it.

Database (as opposed to a simple *data file*) means an organized arrangement of information that can be added to, modified, or queried in various ways that can be automated. For example, suppose you have a long-term study in which people answer questionnaires and respond to questions over a period of several years. You want the database to keep track of participants, to automatically remind them if they have not completed what they should have completed by a certain date, and perhaps to compute scores on the basis of their answers and give feedback or to present selected follow-up questions contingent on a computation based on previous answers.

You might want to allow participants to complete part of a questionnaire and return later to complete the rest of it, and in this case you would want the computer to remind them where they left off. The database can hold information from previous surveys, and the server-side software can inquire of the database whether the participant has completed everything he or she should have done by a given due date (which might depend on the date of the previous participation by the participant). To accomplish these goals, your best solution is to install the Apache server, PHP, and MySQL.

One can find installation packages and instructions on the Web. Göritz has used this method and has written tutorials on how to accomplish many of the useful tasks of longitudinal research. Her teaching materials can be found via the supplementary Web site for the book.

You might wish to keep track of a panel of participants (a subject pool or an online panel). Göritz has written some special programs in PHP that accomplish these tasks, which are free and available on the book's supplementary Web site.

Client-Side Programming

Besides HTML, which is delivered by the server but instructs the client's browser how to display the materials, there are three powerful approaches to programming the client's computer to run a survey or experiment: JavaScript, Java, and Authorware. Because these programs run on the client's side, the client must have something installed and turned on for them to work.

JAVASCRIPT

The first of these languages is JavaScript, which can be used to verify a participant's response, measure response times, randomize assignment to conditions, and do other tasks that require interaction with the participant. This language is supported by most browsers and is now used throughout the Web. In its early days, some users turned it off out of fear of the security lapse of allowing somebody else to run a program on their machines. Schwarz and Reips (2001) noted that when a task can be done equally well by the server or on the client's machine, it may be preferable to use the server-side program rather than lose people who do not have JavaScript installed. Of these three programming approaches, JavaScript is probably the one most likely to run on your participant's machine.

An introduction to basic JavaScript is given in Birnbaum (2001), who provided three chapters on this approach. There are many tutorials on the Web and books on the topic. Birnbaum and Wakcher (2002) showed how the language can be used to control an experiment on probability learning and present a basic tutorial with a series of examples, which are available at the supplementary Web site for this chapter.

JavaScript can be used in very small bits and pieces to perform simple tasks such as randomization of order of presentation of materials, random assignment of people to conditions, checking for reasonable responses or missing responses, measuring response time, and many other such tasks. For more information on JavaScript, visit the following URL: http://www.webreference.com/js/resources/.

JAVA

Java is a powerful object-oriented programming language. It should not be confused with JavaScript, even though these two languages have some similarities. Whereas JavaScript is typically included in a Web page in script form, programs written in Java are precompiled and saved as byte codes that are delivered to the client as *applets,* much like photographs or other media. The client must have the Java engine installed and turned on for the applet to work. The language allows good control of graphics, timing control, and time measurement. Java is a good approach for a person who is a good computer programmer or who is willing to put in the work to become one. Some illustrations of what can be done via Java are provided in McClelland's (2000) *Seeing Statistics:* http://psych.colorado.edu/~mcclella/java/zcalc.html and http://psych.colorado.edu/~mcclella/ATIJava/.

Francis, Neath, and Surprenant (2000) described how Java can be used to conduct many classic examples of cognitive psychology studies. These demonstrations as well as Java experiments on social psychology are now available from Wadsworth, at http://coglab.wadsworth.com/ and http://soclab.wadsworth.com/.

AUTHORWARE

Many of the same tasks that can be done through Java can also be done by means of the fairly expensive program, Authorware. Programming in Authorware involves pushing icons representing various actions such as timed visual displays, loops, computations, interactions with the user, and so on. A brief description of this approach is available at http://www.apa.org/science/psa/williams_prnt.html. Many studies in cognitive psychology created in Authorware are available from the Ole Miss site, created by McGraw, Tew, and Williams (2000), at the following URLs:

http://psychexps.olemiss.edu/
http://www.psych.uni.edu/psychexperiments/.

The use of Authorware software to run experiments requires that the participant has the Authorware player installed on his or her machine. That means that Authorware is an appropriate choice for experiments that can be done in a lab with computers on which the appropriate player has been installed. Another drawback is that Adobe announced that it will support the existing software but cease development of Authorware.

Besides Java, JavaScript, and Authorware, there are other ways to use the client's computer to control the study. At the present time, I cannot recommend that people use software for this purpose, however. Besides the costs of such commercial software, Microsoft software is known for security problems, bugs, and incompatibilities with computer systems running rivals' software. When you conduct experiments via the WWW, you cannot force your participants to use Microsoft products (unless you run in the lab), so it is a strong limitation to use software that works properly only when your Web participants are using Microsoft products.

Web Research Tools

Reips, in association with several collaborators, has created a number of tools for the Web experimenter. These are available at http://psych-iscience.unizh.ch/index.html.

LogAnalyzer (Reips & Stieger, 2004) is a program for extracting information from server logs. When studies are set up as a factorial design, for example, with different treatment combinations in different Web pages, this program can organize and analyze data for analyses of variance, analyze drop-out rates and response times for each treatment condition, and explore many other aspects of the data.

WEXTOR (Reips & Neuhaus, 2002) helps experimenters create and organize the materials for complex experimental designs that can have within-subject and between-subjects conditions. This program also creates a graphical representation of the experimental or quasi-experimental design. Other resources for online experimentation are also available from the above site, including the opportunity to recruit participants from the Web Experimental Psychology Lab (Reips & Lengler, 2005).

Deciding What You Need

Table 2.2 presents a summary of the major techniques of Web-based research that can help you decide what you need to accomplish your type of research online. As noted earlier, anyone planning to do Web-based research needs to know basic HTML (including the technique of Web forms).

TABLE 2.2

Summary of Major Techniques of Web-Based Behavioral Research

Technique	Uses and considerations
HTML: A free programming language. An HTML file is a plain text file whose name has an extension of .htm or .html (e.g., MyWebPage.htm). You can construct the file in a plain text editor (e.g., Notepad) or in a Web page editor (e.g., Dreamweaver). Do not leave spaces in the names of these files; instead, you can use the underscore character (shift-minus), as in my_web_page.htm. If you plan to do research via the WWW, you need to learn basic HTML.	Basic to Web research. Presents and formats text, pictures, graphics, sounds, video, media. See Birnbaum (2001; chaps. 2, 3, 4). Many free Web sites contain good tutorials and summaries. Simple programming effects can be created by design of hyperlinks. For example, if you have different surveys for men and women, smokers and nonsmokers, and so on, you can use links to branch in a survey.
HTML forms: Part of HTML. The free programs SurveyWiz and FactorWiz create Web pages containing HTML forms that send the data to a generic CGI script, which organizes and saves the data sent via the form, and directs the participant to a thank-you page.	See Birnbaum (2001, chap. 5). See also the Perl script by Schmidt that emulates the generic script used by Birnbaum, called "generic.pl," which works with any of the HTML forms in Birnbaum's (2001) book.
SurveyWiz, FactorWiz: Free programs that allow you to make a simple questionnaire or factorial experiment quickly. They are Web pages that make Web pages that run experiments and surveys on the Web.	See Birnbaum (2000). These are easy to learn and easy to use. They are relatively restricted in what they do. You can mix questions requiring a short typed answer or numerical response, with questions with scales of radio buttons or multiple choice. They also allow blending with graphics to put almost any paper-and-pencil study on the Web.
WWW Survey Assistant: Free software, written by Schmidt (1997), that is more powerful than SurveyWiz, but it requires more effort to learn. Creates both the HTML and the Perl scripts to make computations on the data as well as save them to the server.	See links by Schmidt (1997), (http://www.mohsho.com/s_ware/how.html) which describe comparisons of this free software against commercial products.
WEXTOR: A free program that creates the HTML files needed for all sorts of experimental designs, especially between-subjects designs that use different Web pages for different parts of the study.	See Reips & Neuhaus (2002) and materials by Reips, Blumer and Neuhaus (http://psychwextor.unizh.ch/wextor/en/index.php). Advantage of breaking up a study into smaller parts: You can study dropout in detail. WEXTOR creates a visual display of design and skeleton of the Web pages.
LogAnalyzer: Program that analyzes log files.	This program fits well with the WEXTOR approach, in that one can study all of the requests for files and places where dropouts occur.

(continued)

TABLE 2.2 (*Continued*)

Summary of Major Techniques of Web-Based Behavioral Research

Technique	Uses and considerations
Java: Free programming language that typically runs on client side. Neither you nor participant need buy anything. Participant must have it installed and turned on, which is true for most users.	Very powerful, object-oriented programming language. Can be used to make stand-alone programs or Web applets. Java has other uses as well, including precise control and measurement of graphics and events on the screen, such as the position of the mouse. A possible disadvantage is that Java is not installed client-side for some people.
JavaScript: Free programming language that runs on client side. Neither you nor the participant need buy anything, but participants must have it installed and turned on. The scripts can be included in the Web page, which makes your studies open source, which allows others to replicate exactly and build on your materials.	Powerful language. Can add little bits to Web pages to add functionality. Can make programs, including ones to control experiments, manipulate sequence, randomize, time, measure time, and so on. See Birnbaum (2001; chaps. 17–19). A possible disadvantage is that JavaScript is not installed client-side for some people.
Authorware: A commercial program that allows construction of dynamic experiments that run on the client's side. This approach is expensive has great power in creating experiments with control of timing, randomization, detection of screen events, insertion of media, and many other features. The participant must have installed the Authorware Player for the experiments to work. Best in the lab, or with an online panel of participants who have agreed to participate (e.g., they are paid) and have installed the player.	This technique can do many of the same things as Java but uses a graphic user interface with icons to control the experiment (McGraw, Tew, & Williams, 2000). The Ole Miss site (http://psychexps.olemiss.edu/) uses this method for its main power, but this approach also uses HTML and JavaScript as well as server-side programming, Excel macros, and other techniques. Study the manuals that come with Authorware program. A possible disadvantage is that Authorware player is not installed client-side for some people.
Server-side programming: Programming ran on your own Web server, including having CGI scripts installed to save data. Install script on your own server to save to a secure location. This works for all people.	See examples and materials by Göritz (2004). Server-side programming can also be used to do certain other tasks besides saving data (e.g., random assignment to conditions). This is the only method for anything requiring security, such as scoring an exam online or using passwords to sign in to the system.
Apache HTTP Server: Free software server. Göritz (2004) explained how to install this powerful program, which allows you to manage and run your own Web site(s). This gives you complete control over your experiments and data.	Running your own server might be an extra burden on your time and energy. For example, you may need to restart your machine after a system crash or power outage, which may be difficult if you travel a lot.

TABLE 2.2 (*Continued*)

Summary of Major Techniques of Web-Based Behavioral Research

Technique	Uses and considerations
Perl: A free programming language that can be used for server-side programming.	Perl can be used to write CGI scripts that save data from your survey or experiment on your server, for example.
PHP: A free method for server-side programming.	Göritz (2004) described this technique, which has become very popular. Perl and PHP can do many of the same tasks. See also Göritz and Birnbaum (2005).
MySQL: A package for database management, which can be dynamically linked to Web content. For example, one could compute statistics from a study and program it to automatically update online as more people participate. This package is also free.	Göritz (2004) discussed how to install and use MySQL. For some applications, you need only a simple script to save data in a file. However, for many dynamic tasks, the database can provide the solution. One can keep track of people who come and go to the site, remind them what they have and have not finished, and so on.

A person who plans to conduct surveys and Web-based experiments that are of the type that used to be done by pencil and paper can work effectively without needing to know much about how to operate or program a server. Using a server that is operated by a university department or a commercial service provider, one can conduct such research using a generic CGI script to save the data and redirect the participant to an appropriate page. The generic Perl script or the generic PHP forms processor can be used, along with tools such as SurveyWiz and FactorWiz or WWW Survey Assistant to create HTML pages that work with these resources.

Many tasks can be done either by client-side programs or server-side programs: random assignment to conditions, random ordering of the stimuli or items in a questionnaire, timing of displays, response time measurement, and interaction with the participant. Other tasks can be done only by server-side programming. For example, the saving of data to the server can be done only on the server side. Anything requiring the use of passwords, exam keys, and similar issues that require real security should be done only on the server side.

The best choices for free, client-side software are JavaScript and Java. Studies involving randomization of trials, probabilistic stimuli, probabilistic reinforcement, or control of temporal presentations can be done by either of these programming languages. For fine control of visual displays, Java is probably the better choice, although the programming language can be more difficult to learn. Authorware is not free, but it allows programming by moving icons on a flowchart. Java programming is done through lines of text. Java applets are usually precompiled and sent to the participant's computer in the form of byte codes.

The most popular choices for free server-side software are Apache for the Web server, Perl or PHP for the CGI programming, and MySQL for the database software (Göritz, 2004; in press). These software products are open source and free. These techniques are needed when you plan to manage panels of participants, allow participants to interact with each other, keep track of performance by participants over a long period of time, and perform other such dynamic and interactive tasks.

References

Birnbaum, M. H. (Ed.). (2000a). *Psychological experiments on the Internet.* San Diego, CA: Academic Press.

Birnbaum, M. H. (2000b). SurveyWiz and FactorWiz: JavaScript Web pages that make HTML forms for research on the Internet. *Behavior Research Methods, Instruments, & Computers, 32,* 339–346.

Birnbaum, M. H. (2001). *Introduction to behavioral research on the Internet.* Upper Saddle River, NJ: Prentice Hall.

Birnbaum, M. H., & Reips, U.-D. (2005). Behavioral research and data collection via the Internet. In R. W. Proctor & K.-P. L. Vu (Eds.), *Handbook of human factors in Web design* (pp. 471–491). Mahwah, NJ: Erlbaum.

Birnbaum, M. H., & Wakcher, S. V. (2002). Web-based experiments controlled by JavaScript: An example from probability learning. *Behavior Research Methods, Instruments, & Computers, 34,* 189–199.

Fraley, R. C. (2004). *How to conduct behavioral research over the Internet: A beginner's guide to HTML and CGI/Perl.* New York: Guilford Press.

Francis, G., Neath, I., & Surprenant, A. M. (2000). The cognitive psychology online laboratory. In M. H. Birnbaum (Ed.), *Psychological experiments on the Internet* (pp. 267–283). San Diego, CA: Academic Press.

Göritz, A. S. (2004). Apache, MySQL, and PHP for Web surveys. http://www.goeritz.net/ati/ [WWW Document.]

Göritz, A. S. (in press). Building and managing an online panel with phpPanelAdmin. *Behavior Research Methods.*

Göritz, A. S., & Birnbaum, M. H. (2005). Generic HTML form processor: A versatile PHP script to save Web-collected data into a MySQL database. *Behavior Research Methods, 37,* 703–710.

Krantz, J. H. (2001). Stimulus delivery on the Web: What can be presented when calibration isn't possible? In U.-D. Reips & M. Bosnjak

(Eds.), *Dimensions of Internet science* (pp. 113–130). Lengerich, Germany: Pabst Science.

Krantz, J. H., & Dalal, R. (2000). Validity of Web-based psychological research. In M. H. Birnbaum (Eds.), *Psychological experiments on the Internet* (pp. 35–60). San Diego, CA: Academic Press.

McClelland, G. (2000). *Seeing statistics.* Duxbury Press.

McGraw, K. O., Tew, M. D., & Williams, J. E. (2000). PsychExps: An online psychology laboratory. In M. H. Birnbaum (Eds.), *Psychological experiments on the Internet.* (pp. 219–233). San Diego, CA: Academic Press.

Reips, U.-D., Blumer, T., & Neuhaus, C. (2009). WeXtor 2.5: Develop, manage, and visualize experimental designs and procedures. [WWW document]. http://psych-wextor.unizh.ch/wextor/en/index.php

Reips, U.-D., & Lengler, R. (2005). The *Web experiment list:* A Web service for the recruitment of participants and archiving of Internet-based experiments. *Behavior Research Methods, 37,* 287–292.

Reips, U.-D., & Neuhaus, C. (2002). WEXTOR: A Web-based tool for generating and visualizing experimental designs and procedures. *Behavior Research Methods, Instruments, & Computers, 34,* 234–240.

Reips, U.-D., & Stieger, S. (2004). Scientific LogAnalyzer: A Web-based tool for analyses of server log files in psychological research. *Behavior Research Methods, Instruments, & Computers, 36,* 304–311.

Schmidt, W. C. (1997). World-Wide Web survey research made easy with WWW Survey Assistant. *Behavior Research Methods, Instruments, & Computers, 29,* 303–304.

Schwartz, A. (1998). Tutorial: PERL, a psychologically efficient reformatting language. *Behavior Research Methods, Instruments, & Computers, 30,* 605–609.

Schwarz, S., & Reips, U.-D. (2001). CGI versus JavaScript: A Web experiment on the reversed hindsight bias. In U.-D. Reips & M. Bosnjak (Eds.), *Dimensions of Internet science* (pp. 75–90). Lengerich, Germany: Pabst Science Publishers.

CONSIDERATIONS WHEN DESIGNING WEB PAGES | II

Ulf-Dietrich Reips

Design and Formatting in Internet-Based Research

3

The enormous growth in Internet-based research over the past decade has brought with it numerous questions about the most effective way to administer surveys and experiments via the Web. Some research has already been done examining seemingly minor but methodologically substantial issues, such as the relative effectiveness of different presentation formats. Formats can have an effect on sampling: Buchanan and Reips (2001), for instance, found personality differences between Macintosh and PC users, and educational differences between those who had JavaScript turned on in their Web browsers and those who did not.

One of the few disadvantages of Internet-based research is the difficulty of ensuring understanding of the instructions and materials with remote participants (Reips, 2002c), so every effort should be taken to ensure that Internet-based studies are designed most appropriately to fulfill this prerequisite. The present chapter summarizes what can be done and covers basic information that helps researchers

- become sensitive to design and formatting issues and identify how they may play a role in their Internet-based research and
- find out how to best create and run their own Web studies in a way that design and formatting do not interfere with the research but rather support its cause.

This chapter will thus be useful to behavioral scientists who are considering Internet-based data collection as part of their research strategy or as a topic of their teaching. It will also be useful to those who are considering setting up traditional laboratory studies with Internet technologies or are creating Web pages to interact with users, clients, or students in any other ways.

Researchers who read this chapter will be able to identify how and why a certain way of designing a Web study may be a useful method for their research and what to look out for when choosing a different format. It will also help in analyzing work by others, as a reviewer of a paper or grant proposal, for example, or as a teacher, and will help in pinpointing likely problems with the design.

Examples of Design and Formatting Issues

Everyday design and formatting issues can be observed in questionnaires on the Internet. Basically, the issues can be grouped into blatant errors and design decisions made consciously but without the designer realizing that the format makes the site less useable or worthless for research purposes. Figure 3.1—a screenshot from a real student project survey submitted to the Web survey list at http://wexlist.net/—shows several blatant errors in survey design and use of Web forms. These are, in order of severity:

- preselected answers in all drop-down menus (see discussion later in the chapter),
- overlapping answer categories (e.g., which option to choose if one regularly has online contact with 20 online friends?),
- size of text to be entered in text fields is not limited (I entered series of 9s in some cases to demonstrate the problem),
- lack of options that indicate reluctance to answer (e.g., "don't want to answer"),
- all items on one run-on Web page (see discussion below), and
- incorrect writing (e.g., missing comma in the first sentence, other punctuation mistakes, repeated words, and confusing structure of the third item in "Social Networking").

Furthermore, the URL of the survey (not shown) includes the word *student* and thus may (correctly) convey the impression that the survey is part of a student project, thereby not requiring the same attention as what the potential participant may deem "serious research."

When designing a Web study, one is repeatedly confronted with decisions. How many question items may a study have on the Internet? Should question items be all on one page, chunked into groups of items,

FIGURE 3.1

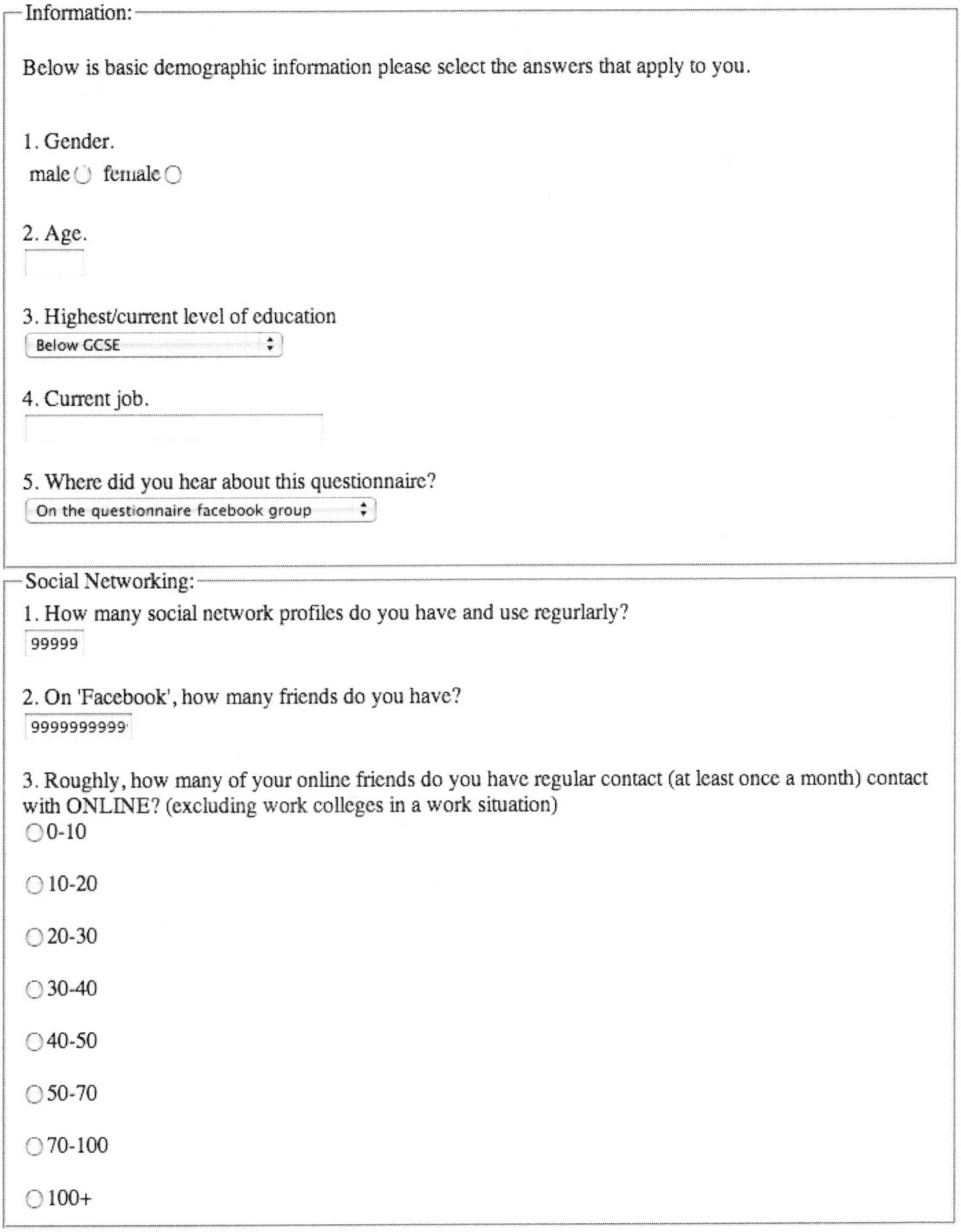

Information:

Below is basic demographic information please select the answers that apply to you.

1. Gender.
   male ○ female ○

2. Age.
   [   ]

3. Highest/current level of education
   [Below GCSE   ▼]

4. Current job.
   [   ]

5. Where did you hear about this questionnaire?
   [On the questionnaire facebook group   ▼]

Social Networking:

1. How many social network profiles do you have and use regurlarly?
   [99999]

2. On 'Facebook', how many friends do you have?
   [9999999999]

3. Roughly, how many of your online friends do you have regular contact (at least once a month) contact with ONLINE? (excluding work colleges in a work situation)
   ○ 0-10

   ○ 10-20

   ○ 20-30

   ○ 30-40

   ○ 40-50

   ○ 50-70

   ○ 70-100

   ○ 100+

Example portion of a questionnaire that recreates a student research project on the Web. It shows several errors in study design and use of form elements that will inevitably result in biased data.

or in a one-item–one-page format? What types of (new) response options are there on the Web, and how do they work? What are the pitfalls of using them? What will happen if you tell participants at the beginning of the study how long it will take, and what if you do not? What if you provide feedback on respondents' progress during the study (e.g., by using a progress bar)? Where should demographic questions be placed, at the beginning or at the end, and how will this affect data quality? What are hidden formatting issues? How much of an issue is order of items or pages for Internet-based research? Once you know what you want to do, how do you create and edit the pages—Which software is available (for free?) and which one best suits your purposes? After reading this chapter you will know how to answer these questions.

Understanding Why Web Design and Format Are Important for Internet-Based Research

At the core of many of the more important methodological problems with design and formatting in Internet-based research are *interactions between psychological processes in Internet use and the widely varying technical context* (Krantz, 2001; Reips, 2000, 2007; Schmidt, 2007). Data quality can be influenced by degree of anonymity, and this factor as well as information about incentives also influences the frequency of drop out (Frick, Bächtiger, & Reips, 2001; O'Neil & Penrod, 2001). The degree of personalization and the power attributable to the sender of an e-mailed invitation to participate in an Internet-mediated study affect the response rates (Joinson & Reips, 2007). Dillman and colleagues (Dillman & Bowker, 2001; Smyth, Dillman, & Christian, 2007; Smyth, Dillman, Christian, & Stern, 2006) have shown that many Web surveys are plagued by problems of usability, display, sampling, or technology. Design factors such as the decision whether a one item, one screen (OIOS) procedure is applied may trigger context effects that lead to results differing from those acquired with all questions on one Web page (Reips, 2002a, 2007).

DESIGN FACTORS

Design may explain differences between results from online and offline methods (for mixed modes, see Shih & Fan, 2008), and researchers are trying to find ways of bringing established offline methods to the Internet (e.g., to recreate the personal presence of an interviewer by using videos in online surveys; Fuchs, 2009). Small changes in design may cumulatively have large effects on participants' decisions to provide data or not. For example, Reips (2000) listed various design factors as part of the *high entrance barrier* or *high-hurdle technique,* a package of procedures that can be applied to provoke early drop out and trigger compliance after someone makes the decision to participate (for a detailed explanation, see chap. 13, this volume). Several factors that may lead to a participant keeping motivation to be in the study are often placed at the beginning of a Web experiment (i.e., on the general instructions page), for instance, when researchers

- "tell participants participation is serious, and that science needs good data;
- personalize the research (e.g., by asking for e-mail addresses, phone number, or both);
- tell participants they are traceable (via their computer's IP address);
- are credible (e.g., by telling participants who the researchers are and what their institutional affiliation is);
- tell participants how long the Web experiment will take;
- prepare participants for any sensitive aspects of the experiment (e.g., 'you will be asked about your financial situation');
- tell participants what software they will need (and provide them with hyperlinks to get it);
- perform Java, JavaScript, and plug-in tests" (Reips, 2000, pp. 110–111); and
- ensure that the technique can be supported by adjunct procedures (e.g., a Web design that results in systematic shortening of loading times).

THE ONE ITEM, ONE SCREEN, DESIGN STRATEGY AND OTHER ISSUES OF GROUPING AND LENGTH

Even for very long studies with several dozens or several hundred items, you will be able to find participants on the Internet (see, e.g., chaps. 10 and 11, this volume). Generally, however, about 25 to 30 questions of medium complexity should be the upper limit (Gräf, 2002; Krasilovsky, 1996; Tuten, Urban, & Bosnjak, 2002). *Medium complexity* means a level of complexity between single-item, two-choice questions (i.e., not complex) and matrix question or lengthy open-ended questions (i.e., highly complex).

Items can be grouped in thematic chunks of three to four on a single screen. However, there are several good reasons to always consider the OIOS strategy:

- Context effects (interference between items) are reduced (Reips, 2002a),

- meaningful response times and drop out can be measured and used as dependent variables, and
- the frequency of interferences in hidden formatting is vastly reduced (for a clarifying example with radio buttons named "sex," see Reips, 2002b).

Crawford, Couper, and Lamias (2001), among others, investigated the effects of telling participants at the beginning of a Web study about its duration. If a longer time was given, then fewer persons began the study. However, once they had decided to participate, more finished the study than those in a group that was given a lower duration. Similarly confirming the high-hurdle idea, Ganassali (2008) concluded, "We can say that the decision to quit the survey is influenced by perceived length and by style of wording, on the very first pages of the form." She found experimental evidence for positive effects of short, direct, and interactive formats in Web surveys. Progress indicators appeared to have a negative effect on completion rate in Crawford et al.'s study. More recent studies have indicated that there may be an interaction with length of study: Progress indicators have a motivating effect in short studies and are demotivating in long studies (see also chap. 15, this volume).

Frick, Bächtiger, and Reips (2001) showed lower drop out and better data quality, that is, significantly fewer missing data, for early versus late placement of demographic questions. They also manipulated whether information about an incentive for participation in the experiment was given at the beginning or the end. Drop out was 5.7% when both incentive information and demographic items were given at the beginning and was 21.9% when both were given at the end, with the other two conditions in between: 13.2% when incentive information was given at the beginning and demographic items at the end, and 14.9% vice versa.

A WORD ON (HYPER)TEXT

Text is more difficult to read on screens than on paper, and handling online documents involves more behaviors than opening books and flipping pages. Thus, keep in mind that Internet participants will more quickly reach their thresholds for perceived burden and attention capacity, and their cognitive processes may more easily get distracted. Studies on knowledge acquisition with hypertexts show that a linear static format better supports understanding for single texts than does an active presentation, in which users have to decide themselves where to move and then scroll or click to do so (Naumann, Waniek, & Krems, 2001). So you fare best to use the linear static format as the default, because most Internet studies are of that type. However, for multiple hypertexts, the construction of a mental model is moderated by task: Active presentation supports understanding in argumentative tasks and static presentation format does so for narrative tasks (Hemmerich & Wiley, 2002; Wiley, 2001). Accordingly, format your study materials in line with the tasks you are using.

Formatting Issues in Study Design

In this section, I cover issues in formatting of Web studies, namely, how to avoid pitfalls in creating page titles, page names, and Web forms, and how to use hidden formatting, and response options. Note that many of the recommendations are followed by some, but by far not all software that generates Internet-based studies: Check before you buy.

TITLES AND NAMES

Generally, use an informative title on Web pages announcing the study and on the first page of the study to attract participants. For subsequent pages, use uninformative page titles so people will not find these pages via search engines and enter the study somewhere in the middle.

HIDDEN FORMATTING

Remember that Web page design is a bit like carving icebergs: Much is happening below the surface. Hidden formatting can become very complex and complicated. *Server-* and *client-side scripting* and *server-side includes*[1] are powerful tools that allow Web designers to achieve wonderful things if everything works well. In the interest of space, I do not discuss those advanced possibilities here; for more information on this, the interested reader can read related chapters in this volume and the additional resources at the end of this chapter. Here, I will describe some basics of hidden formatting for illustration.

As an example of the power of hidden formatting, consider open questions. These are created by using text area or text field tags. Because just one of your Internet participants may, for example, copy and paste the entire text of *Harry Potter and the Goblet of Fire* into your carefully formulated text field to crash your server or overwrite your data file, you need to limit the number of characters that can be pasted into text fields.

---

[1]For more information on server-side includes see http://en.wikipedia.org/wiki/Server_Side_Includes

This is done by including a hidden "maxlength" attribute with the text field tag, for example:

<input type = "text" name= "myopenfield1" size="24"
maxlength="30" border="0">.

Here, the text field will be displayed as 24 characters long (i.e., size), but it will take up to 30 characters (i.e., maxlength).

An important technique in Internet-based research is the use of *meta tags*. Meta tags serve a variety of purposes, for example they keep search engines away from all pages except the first page (so participants do not enter the study on one of the later pages), and they prevent the study materials from being cached. Meta tags are hidden in a Web page's "head" section and look as follows on WEXTOR-generated (http://wextor.org; Reips & Neuhaus, 2000) pages:

<meta name="ROBOTS" content="NONE">

<meta http-equiv="pragma" content="no-cache">

<meta http-equiv="expires" content="Thursday, 1-Jan-1991
01:01:01 GMT">

<meta http-equiv="content-type" content="text/html;
charset=ISO-8859-1">

The "ROBOTS" tag is set to "NONE," because the routines used by search engines to search the Web for new Web pages are named *robots* (sometimes *spiders* or *crawlers*). Thus, the ROBOTS tag informs the routines there is nothing to be catalogued. The two meta tags that follow prevent caches in mediating servers, search engines, and proxy servers from serving old versions of research materials after they have been updated. Caches contain stored files downloaded from the Web, for later reuse. Internet providers and large institutions run computers with large hard disks (mediating servers or proxy servers) to store hypertext markup language (HTML) code, images, and other media. If a page is requested again from within their network, the server quickly checks in the cache if it holds any of the text and media and sends it instead of letting the request go all the way out to the Internet to retrieve the original material. This way, the page can be displayed more quickly, and much unnecessary traffic is avoided. However, the material loaded from the cache may be outdated: If an experimenter finds an error in the material and replaces the Web page, users may continue to be delivered the old version. The meta tags displayed above will prevent this.

Participants may search the Web using *keywords* and thus find studies that use these terms in a keyword meta tag on the first page. Keywords may be a good way to recruit participants with a particular interest or for long-term Internet-based studies (Birnbaum & Reips, 2005).

CONFIGURATION ERRORS: PITFALLS IN WEB FORMS

Reips (2002b) discussed how to avoid several potential configuration errors with Web study design: Protection of directories, a suitable data transmission procedure, unobtrusive naming of files and conditions, adaptation to the substantial variance of technologies and appearances of Web pages in different browsers, and proper form design (Configuration Errors I–V). Another type of configuration error involves Web forms. Web forms that were available since the World Wide Web consortium (W3C) announced the standard HTML 2.0 in 1995 (Musch & Reips, 2000) constitute much of what can be done interactively on Web pages. Mastering the design of Web forms can be tricky, as can easily be observed on the Web. Figure 3.2 from Reips (2009) shows several problems with a large publishing company's feedback form:

- Preselected answers (Questions 5 and 7): Skipping the question will show up as having selected the preset option;
- Participant burden (Question 5): Selecting an option other than the default is highly discouraged by signaling further work to the respondent ("Please provide address"); and
- Mandatory responses (Question 6): Even though Question 6 will often need no answer, the survey designers set the field to require a response (for example, a dash), as indicated by "*".

For more details on configuration errors in Internet research, see chapter 13 of this volume.

RESPONSE OPTIONS

All traditional response option formats have been researched for use on the Web (e.g., open-ended formats versus close-ended formats; Reja, Lozar Manfreda, Hlebec & Vehovar, 2003). Here, I focus on new types of response options. With these, Internet-based research offers types of dependent variables that are not available in paper-and-pencil research and are impractical in offline computer-based research, such as drop-down menus (selection lists) and visual analogue scales (VAS).

Drop-down menus are manually complex response devices. To use them, participants have to click on the one option that is initially displayed. Then, once it expands to show other alternatives, they have to either scroll and click or drag a certain distance and then release at a certain choice, depending on their operating system. Reips (2002b; Configuration Error V) emphasized the importance of not having a legitimate response preselected, so real answers can later be distinguished from failures to respond. Because of the complexities of the device, Birnbaum and

FIGURE 3.2

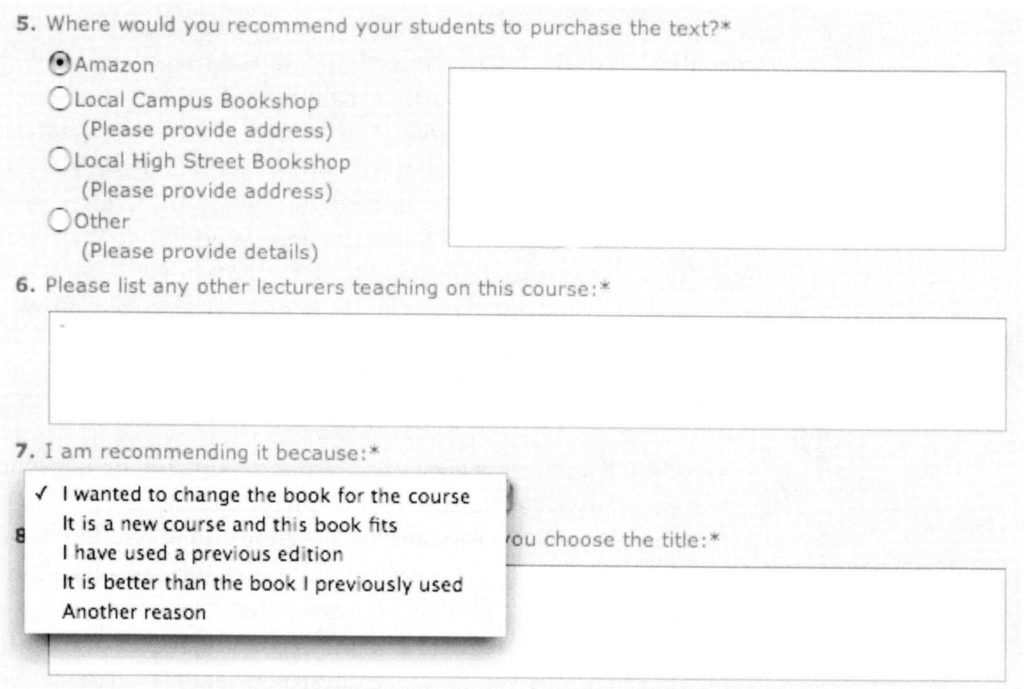

5. Where would you recommend your students to purchase the text?*

- ◉ Amazon
- ○ Local Campus Bookshop
  (Please provide address)
- ○ Local High Street Bookshop
  (Please provide address)
- ○ Other
  (Please provide details)

6. Please list any other lecturers teaching on this course:*

7. I am recommending it because:*

| ✓ I wanted to change the book for the course |
| It is a new course and this book fits |
| I have used a previous edition |
| It is better than the book I previously used |
| Another reason |

8.   ou choose the title:*

Improper use of Web form elements that results in biased data. The figure is adapted from Reips (2009) and shows a portion of a publisher's online questionnaire for feedback on book inspection copies.

Reips (2005) thus recommended precaution with drop-down menus and warned of another potential issue with them:

> Another problem can occur if the value used for missing data is the same as a code used for real data. For example, the first author found a survey on the Web in which the participants were asked to identify their nationalities. He noted that the same code value (99) was assigned to India as to the preselected "missing" value. Fortunately, the investigator was warned and fixed this problem before much data had been collected. Otherwise, the researcher might have concluded that there had been a surprisingly large number of participants from India. (p. 474)

VASs markedly show how previously known advantages of a measure that was often not used for practical reasons (e.g., burdensome measurement by hand) in offline environments becomes highly valuable when taken online. Reips and Funke (2008) developed VAS Generator (http://vasgenerator.net), a free tool to create VAS for Internet-based research. Their research shows that the online VAS created with the tool produce data on the level of an interval scale, even for extreme scale lengths of 50 and 800 pixels.

The abundance of further issues with these and other types of response options cannot be covered here. Also, much has not yet been researched or will need to be researched again because the Internet environment is constantly changing.

Formatting Companions: Page Editors

In Table 3.1, I provide an overview of program and editor options to create and edit Web pages. Even if you use a fully automated Web application for study design like WEXTOR (http://wextor.org), Surveymonkey (http://surveymonkey.com), or Unipark (http://unipark.com), you will need one of these to check and understand your study's format and make minor edits.

The selection displayed and commented on in Table 3.1 is subjective, and many more editors are available for different operating systems. However, this selection was agreed upon by several instructors at the National Science Foundation's and American Psychological Association's Advanced Training Institutes "Performing Web-Based Research."

Conclusion

The present chapter will help you in designing and formatting your Internet-based research. Be aware, though, that many factors interact with each other, and both Web technology and user experience are constantly changing. More research is needed in light of the rapid development in Web technology, changes in user experience, and the consequences of formatting and design in Internet-based research.

Additional Resources

iScience Server: http://www.iscience.eu/.
A portal with a number of useful services in Internet-based research.

TABLE 3.1

Recommendations for the Use of HTML Editors

Software, operating system, and availability	Description
Notepad Windows Free	Text only. Allows direct editing of HTML. Be careful saving the file or it will add a .txt extension.
BBEdit Lite (now TextWrangler) Mac OS, Mac OS X Free, http://www.barebones.com/products/TextWrangler/download.html; commercial version is more powerful	A code editor that will highlight syntax like HTML screens of WYSIWYG editors and WinEdit.
Word Windows and Mac Commercial	Can create Web pages but adds a lot of Microsoft-specific style information that does not do well in some browsers.
Dreamweaver Windows and Mac Commercial (free trialware; http://www.adobe.com/go/trydreamweaver ~$200, educational discount)	Commercial WYSIWYG editor. Allows direct editing of HTML.
Expression Web Windows Commercial (free trialware, http://www.microsoft.com/Expression/products/download.aspx?key=web)	Same as Dreamweaver but some functions require usage of server with special extensions. A special plugin ("Silverlight") is required.
UltraEdit Windows Commercial (free trialware, http://www.ultraedit.com/ $79)	Allows direct editing of HTML and other code (not WYSIWYG).
KompoZer Windows, Mac OS X, Linux Donationware (http://kompozer.net/)	An open-source WYSIWYG editor. Allows direct editing of HTML. A simple version of Dreamweaver and Expression Web.

Note. WYSIWIG = what you see is what you get.

Proctor, R. W., & Vu, K.-P. L. (Eds.) (2005). *The handbook of human factors in Web design.* Mahwah, NJ: Erlbaum.
The book provides much practical and useful advice in Web design.

References

Birnbaum, M. H., & Reips, U.-D. (2005). Behavioral research and data collection via the Internet. In R. W. Proctor & K.-P. L. Vu (Eds.), *The handbook of human factors in Web design* (pp. 471–492). Mahwah, NJ: Erlbaum.

Buchanan, T., & Reips, U.-D. (2001, October 10). Platform-dependent biases in Online research: Do Mac users really think different? In K. J. Jonas, P. Breuer, B. Schauenburg, & M. Boos (Eds.), *Perspectives on Internet research: Concepts and methods.* Retrieved December 27, 2001, from http://server3.uni-psych.gwdg.de/gor/contrib/buchanan-tom

Crawford, S. D., Couper, M. P., & Lamias, M. J. (2001). Web surveys: Perceptions of burden. *Social Science Computer Review, 19,* 146–162.

Dillman, D. A., & Bowker, D. K. (2001). The Web questionnaire challenge to survey methodologists. In U.-D. Reips & M. Bosnjak (Eds.), *Dimensions of Internet science* (pp. 159–178). Lengerich, Germany: Pabst Science.

Frick, A., Bächtiger, M. T., & Reips, U.-D. (2001). Financial incentives, personal information, and drop out in online studies. In U.-D. Reips & M. Bosnjak (Eds.), *Dimensions of Internet science* (pp. 209–219). Lengerich, Germany: Pabst Science.

Fuchs, M. (2009). Gender-of-interviewer effects in a video-enhanced Web survey: Results from a randomized field experiment. *Social Psychology, 40,* 37–42.

Ganassali, S. (2008). The influence of the design of Web survey questionnaires on the quality of responses. *Survey Research Methods, 2,* 21–32.

Gräf, L. (2002). Assessing Internet questionnaires: The online pretest lab. In B. Batinic, U.-D. Reips, & M. Bosnjak (Eds.), *Online social sciences* (pp. 73–93). Seattle, WA: Hogrefe & Huber.

Hemmerich, J., & Wiley, J. (2002). Do argumentation tasks promote conceptual change about volcanoes? In W. D. Gray & C. Schunn, *Proceedings of the Twenty-Fourth Annual Conference of the Cognitive Science Society* (pp. 453–458). Hillsdale, NJ: Erlbaum.

Joinson, A. N., & Reips, U.-D. (2007). Personalized salutation, power of sender, and response rates to Web-based surveys. *Computers in Human Behavior, 23,* 1372–1383.

Krantz, J. H. (2001). Stimulus delivery on the Web: What can be presented when calibration isn't possible. In U.-D. Reips & M. Bosnjak (Eds.), *Dimensions of Internet science* (pp. 113–130). Lengerich, Germany: Pabst Science.

Krasilovsky, P. (1996), Surveys in cyberspace. *American Demographics,* November/December, 18–22.

Musch, J., & Reips, U.-D. (2000). A brief history of Web experimenting. In M. H. Birnbaum (Ed.), *Psychological experiments on the Internet* (pp. 61–88). San Diego, CA: Academic Press.

Naumann, A., Waniek, J., & Krems, J. F. (2001). Knowledge acquisition, navigation, and eye movements from text and hypertext. In U.-D. Reips & M. Bosnjak (Eds.), *Dimensions of Internet science* (pp. 293–304). Lengerich, Germany: Pabst Science.

O'Neil, K. M., & Penrod, S. D. (2001). Methodological variables in Web-based research that may affect results: Sample type, monetary incentives, and personal information. *Behavior Research Methods, Instruments, and Computers, 33,* 226–233.

Reips, U.-D. (2000). The Web experiment method: Advantages, disadvantages, and solutions. In M. H. Birnbaum (Ed.), *Psychological experiments on the Internet* (pp. 89–114). San Diego, CA: Academic Press.

Reips, U.-D. (2002a). Context effects in Web surveys. In B. Batinic, U.-D. Reips, & M. Bosnjak (Eds.), *Online social sciences* (pp. 69–79). Seattle, WA: Hogrefe & Huber.

Reips, U.-D. (2002b). Internet-based psychological experimenting: Five *do*s and five *don't*s. *Social Science Computer Review, 20,* 241–249.

Reips, U.-D. (2002c). Standards for Internet-based experimenting. *Experimental Psychology, 49,* 243–256.

Reips, U.-D. (2007). The methodology of Internet-based experiments. In A. N. Joinson, K. Y. A. McKenna, T. Postmes, & U.-D. Reips (Eds.), *The Oxford handbook of Internet psychology* (pp. 373–390). Oxford, England: Oxford University Press.

Reips, U.-D. (2009). *Collecting data in surfer's paradise: Internet-mediated research yesterday, now, and tomorrow.* Manuscript submitted for publication.

Reips, U.-D., & Funke, F. (2008). Interval-level measurement with visual analogue scales in Internet-based research: VAS Generator. *Behavior Research Methods, 40,* 699–704.

Reips, U.-D., & Neuhaus, C. (2002). WEXTOR: A Web-based tool for generating and visualizing experimental designs and procedures. *Behavior Research Methods, Instruments, & Computers, 34,* 234–240.

Reja, U., Lozar Manfreda, K., Hlebec, V., & Vehovar, V. (2003). Open-ended vs. close-ended questions in Web questionnaires. *Advances in Methodology and Statistics (Metodoloki zvezki), 19,* 159–177.

Schmidt, W. C. (2007). Technical considerations when implementing online research. In A. N. Joinson, K. Y. A. McKenna, T. Postmes & U.-D. Reips (Eds.), *The Oxford handbook of Internet psychology* (pp. 461–472). Oxford, England: Oxford University Press.

Shih, T.-H., & Fan, X. (2007). Response rates and mode preferences in Web-mail mixed-mode surveys: A meta-analysis. *International Journal of Internet Science, 2,* 59–82.

Smyth, J. D., Dillman, D. A., & Christian, L. M. (2007). Context effects in Internet surveys: New issues and evidence. In A. N. Joinson, K. Y. A. McKenna, T. Postmes, & U.-D. Reips (Eds.), *The Oxford handbook of Internet psychology* (pp. 429–446). Oxford, England: Oxford University Press.

Smyth, J. D., Dillman, D. A., Christian, L. M., & M. J. Stern (2006). Effects of using visual design principles to group response options in Web surveys. *International Journal of Internet Science, 1,* 5–15.

Tuten, T. L., Urban, D. J., & Bosnjak, M. (2002). Internet surveys and data quality: A review. In B. Batinic, U.-D. Reips, & M. Bosnjak (Eds.), *Online social sciences* (pp. 7–26). Seattle, WA: Hogrefe & Huber.

Wiley, J. (2001). Supporting understanding through task and browser design. In J. D. Moore & K. Stenning, *Proceedings of the Twenty-Third Annual Conference of the Cognitive Science Society* (pp. 1136–1143). Hillsdale, NJ: Erlbaum.

John H. Krantz and John E. Williams

Using Graphics, Photographs, and Dynamic Media

4

Since the advent of Mosaic, it has been possible to include images in World Wide Web pages (Levy, 1995). This simple advance in Web browser capability dramatically transformed people's experience of the Web. From a dry academic technology designed to facilitate the critical reading of documents and information, the Web became a means for sharing of information remotely in a visually appealing format. It seems only appropriate that psychological research would seek to take advantage of these ever-expanding capabilities of the Web. Many areas of psychology rely on the use of the stimulus; however, the vast majority of posted psychological research studies have used survey methodology. Approximately 80% of the current studies posted on the *Psychological Research on the Net* Web site (see the supplementary Web site for this chapter) use survey methodology (Krantz, 2007). These studies use the text-presentation ability of the Web and do not take advantage of the media-presentation ability. There are many reasons for the limited usage of media in Web research. One critical reason is the loss of control over the stimulus environment entailed by the use of the Web (Krantz, 2001). Other issues include the increased complexity of the development of the study entailed by the Web and the desire to reach as wide a participant population as possible.

However, many of these issues can be controlled to some extent or even overcome, which is the point of this chapter. This chapter covers several topics. First, the nature of presenting a stimulus over the monitor is discussed. If the problems of the variable display environment are to be addressed, the nature of the hardware on which the stimulus is to be presented needs to be understood. In this section, recommendations for general media development are covered to help minimize the problems of display variation. Finally, solutions for dealing with both pictures and dynamic media are discussed. In each case, the discussion covers issues related to how to best use the media.

The Monitor

More details about the nature of the monitor and how to display stimuli over the Web can be found in Krantz (2000, 2001). Still, a brief discussion of the nature of the monitor will help illustrate some of the complexity and issues surrounding the development and displaying of media on the Web. However, for give a general sense of the issue, consider that media on the Web can look substantially different from one viewing to the next. For example, the type of monitor that might be used can vary. Many types of technologies are available for computer monitors, most commonly the traditional cathode ray tube (CRT) and liquid crystal diode (LCD). The CRT is the traditional, television type of display that has a deep back; the LCD monitors are flatter and are the flat panel increasingly used with computers.

This discussion about the monitor is broken into three sections, one for each of the dimensions of the operation: space, color, and luminance. Each section covers the basic issues and then covers some recommendations. The section ends with some general recommendations about the use of media.

SPACE

One of the most important issues regarding any stimulus is its size. It is the size of the stimulus at the eye that is important. This stimulus size depends on the monitor resolution, monitor size, and viewing distance. As the monitor gets bigger, so will the stimulus. As the resolution gets finer, the smaller the stimulus gets. The farther the participant is from the monitor, the smaller the effective size of the stimulus. Thus, over the Web, it is impossible to know the size of any stimulus for the participant. In many cases, stimulus size is not of critical importance. If the stimulus is fairly simple and the details are not terribly important, the stimulus size probably does not play a large role. Simple line drawing or figures in which the general shape is the most important attribute fall into this situation. However, for more complicated stimuli, size may be critical. Figure 4.1A is an example of how size effects the image of a face. If the general shape of the face is of primary importance, then the image seems relatively little affected as it goes from a large size to one eighth the original size. However, if the image depends on fine details, changes in size could cause the loss of important information (see Figure 4.1B). In this image, the deer are lost in the smallest image, whereas the details are lost even in the middle-sized image. Limiting stimuli to simple images can reduce the impact of stimulus size.

One of the more difficult issues of monitors is the fact that screens are not uniform over their entire surface (Krantz, 2000, 2001). The intensity, contrast, and even colors of stimuli change over the surface. When practical, try to use the entire screen with a pop-up window and limit stimuli to the central region. Pop-up windows can be created with JavaScript routines. If it is not practical or desirable to open a pop-up window, keep image stimuli small so that the image will not cover too large of an area of the screen and thus will be subject to variations in the screen surface within the image (Krantz, 2000, 2001).

CHROMATIC

Color is not always used in stimuli, but when it is, it is generated by differential activation of the three primaries on the monitor surface. The primaries used are usually red, green, and blue. The range of colors produced is limited by the primaries on the monitor surface. The color wheel will serve adequately to illustrate this issue (see Figure 4.2). Figure 4.2 shows an example of a color wheel. The corners of the triangle drawn on the color wheel represent an ideal set of colored primaries. The colors that can be made by a monitor with these three primaries are the colors that are contained by this triangle, called the "color gamut." Considering that actual monitor primaries do not even sit on the outside of the color circle, it is clear that most monitors are quite limited in their color reproduction abilities. For a full treatment of monitors and color reproduction, see Silverstein and Merrifield (1985). The problem is made worse in that primaries change from monitor to monitor and over time on a given monitor. Fortunately, color perception is relative, to a large degree. What that means is that the perception of one color depends not only on the wavelengths from that point but also on the colors around it, to some degree (Land, 1974).

If the monitor was the only light in the room, it is probable that color constancy would compensate for most issues related to color variation on monitors. However, all of the light in the room and light from any open

FIGURE 4.1

A

B

Effect of image size on image quality. A: Different scales of a picture of a face. Notice that even at the smallest size (one eighth the original), nothing of importance seems to be lost from the image. B: An image with a lot of detail. In the image to the right, it is hard to see that there are deer in the image.

FIGURE 4.2

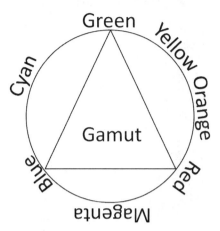

The color wheel, illustrating the color gamut of reproducible colors.

windows is reflected off of the screen surface and adds to the light coming from the monitor. Because the room light and sunlight is usually white, it tends to desaturate, or wash out, the screen. How much the color is washed out depends on the monitor type, the brightness of the room, and the brightness of the monitor. In general, the brighter the room light and the dimmer the color on the monitor, the more it washes out. CRTs reflect the most light, so they are worse than LCDs: the more light in the room also the more washing out. Web experiments with color have revealed some significant problems when run on the Web (Krantz, 2001; Laugwitz, 2001). Figure 4.3 shows the same picture on a CRT both in a dark room (top panel) and a moderately lit room (bottom panel). The effect of room light is clearly shown in the washed-out image in the lit room. As a recommendation, use few colors in your stimulus, a lot of contrast, and highly saturated colors. It is possible that instructions to participants about closing windows and reducing lights might help, but they cannot guarantee compliance.

LUMINANCE

The intensity of each of the primary colors is controlled by a value stored for each pixel. On most monitors, the values are stored in 8 bits ranging from 0 to 255. This makes a total of 24 bits for each pixel on most monitors (although 32 bits per pixel is becoming more common, with the additional 8 bits being used for transparency effects). Many of

FIGURE 4.3

A

B

A photograph displayed on a color monitor in a dark room (A) and in a moderately lit room (B).

the first few bits make no change to the intensity of the image, and after that point, the image intensity is controlled by a function called the "gamma function" (Krantz, 2000). The main point here is that values used to make up the image are not a literal representation of the intensity in the image.

More important to your image is the contrast or difference between adjacent light and dark regions. Being able to resolve or make out an image is in many ways driven more by contrast than by intensity. Given the lighting issues mentioned above and the importance of contrast, maintaining an image of high contrast is very important (refer again to Figure 4.3). Thus, for stimuli, high-contrast images are desirable. Black-and-white images have more contrast than color images because of the use of all of the primaries together creating a greater range of intensity in the images. As a result, black-and-white stimuli are less washed out by lighting than color images. Thus, for photographs, unless the image is of a very high intrinsic contrast, and most photographs are not, converting the photograph to black and white will improve its use on the Web.

GENERAL MONITOR RECOMMENDATIONS

Given the variation of the monitor alluded to so far, careless use of stimuli could lead to artifacts and confounds that can obscure the impact of the variables of interest in a psychological study. However, using simple stimuli that are high contrast seems to be safe on the Web. If more complicated stimuli are needed, the use of black-and-white images is recommended. In any case, it is advisable to pilot test the stimuli using a variety of computers, monitor resolutions, viewing distances, and lighting environments to see how the stimuli are affected by the variations possible on the Web.

Graphics and Still Images

With that basic introduction about the monitor, it is easier to explain what is going on with images, how to edit them, and why the editing operations work the way they do. First, the most important types of image files for Web use are discussed and then how to edit them using the freeware program ImageJ is presented.

TYPES

Three main types of image files need to be understood. The first is a bitmap, or raw data, version. These files have extensions like .tiff, .tif or .bmp (on Windows). They contain a complete set of information to present the image on the screen. If you have a 24 bit 640 × 400 pixel image, you will have a file that will have 24 bits of color information (8 each for red, green, and blue) at each of the 640 × 400 pixels. That is 6,144,000 bits of information for the image, or 6,000 kb, or ~5.9 megabytes. That is a lot of information and can lead to slow download. As a result, most images are compressed.

However, it is important to know that whenever you display or edit any image, even if stored in a compressed format, the image is in a bitmap form. To display any of the compressed forms of image files during, say, editing, the image must be converted to a bitmap form. To then restore the image, it is reconverted back to the compressed form, which can cause further loss of the image information. Thus, sometimes, unpredictable problems can happen when you reopen an image because what is stored is different than what is being edited. As a result, always save a copy of any original image and always edit a new copy of the original image.

Graphics Interchange Format (GIF) is a format for color images developed by the old CompuServe online service using an image compression routine developed by Unisys. The image compression routine is copyrighted: Although the images generated by this routine are free to be used, the software that does the compression needs to have paid a fee to use the routine. Most commercial software packages have licenses to use this routine, so they are safe to use; however, some freeware and other packages should be looked at carefully before generating images with them.

The compression routine for GIF images compresses the color space from 24 bits into 8 bits. It uses a look-up table (LUT) to reproduce the colors in the image. Thus, the spatial information is always preserved, but color information is lost. If the image is a simple graphic that does not use a wide range of colors, the GIF format is an excellent choice. For a photograph that uses a wide range of colors that can exceed GIF's color range, you can lose a lot of image quality and get some strange artifacts.

The other primary image compression format is JPEG. It is a free, or open, format so any one can develop a program to use this format and you do not have to worry about violating a patent or copyright. JPEG does not compress in the color dimension. If the image starts with 24 bits of color, it ends with 24 bits of color; it compresses in the spatial domain. Because of the preservation of total color information, it works very well for photographs, and most digital cameras use the JPEG format as the default. However, the limitation of the JPEG format is that repeated saving will further corrupt the image. This fact is another reason to save the original image and edit from the original image instead of continually going back to copies.

USING IMAGEJ

Many high-quality image editing programs are available, and operating systems are coming endowed with increased image-editing capability. However, many of the commercial image-editing programs are costly. I describe one of these, ImageJ a freeware program developed by Rashband (2003) of the National Institutes of Health.

Description and Installation

ImageJ is a scientific-grade image-editing program used for medical, chemical, and even neuroscience images, for which the need for high-quality image processing is a must. The program is written in Java, so it can be run on any type of computer. You can download it from http://rsb.info.nih.gov/ij/index.html, the home page of the effort to develop this program. The download page has install packages that make program installation easy. After I describe some basic image-editing tasks, I move on to discuss some important ways to customize the program to increase ease of use and power.

ImageJ Interface

Here I describe the basic layout of the program interface (see Figure 4.4). ImageJ is a windowing program so that its controls are structured like most other programs with both menus and toolbars. The menus are across the top and are "File," "Edit," "Image," "Process," "Analyze," "Plugins," "Window," and "Help." The "File," "Edit," "Window," and "Help" menus are similar to the menus of the same name on other programs. In the "File" menu, you open, close, and save the images that you are working on. In the "Edit" menu, you can do some basic editing tasks like cutting and pasting and edit some of the basic options of the program. Also, the basic drawing commands are found in this menu ("Draw" and "Fill").

FIGURE 4.4

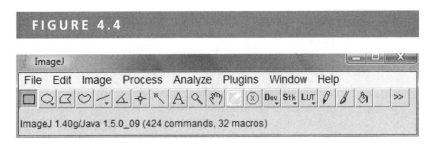

The interface for ImageJ when you open the program.

In the "Window" menu, you can select the window you want to work on. Each image is opened in a separate window, and any dialog boxes that go along with a command will also be in a separate window. The "Help" menu allows you to access the help information, much of which is online. The "Image" menu allows basic image manipulation such as changing image size and cropping. The "Process" menu contains more advanced image manipulation and filtering to smooth or sharpen the image. The "Analyze" menu collects statistics about the image; I do not discuss this menu in more detail, but it is very important to many of the scientific imaging uses of this program, and this is where you obtain your data. The "Plugins" menu is where you can create, edit, compile, and access the plugins that can be added to this program.

The toolbar below the menus starts out from the left with various shapes: a rectangle, oval, polygon, a closed freehand shape, line, a multi-segment open line, and an open freehand. These shapes are used for selections and drawing. The next two buttons on the tool bar are the crosshairs used for analysis and the "magic wand," which does selection based on regions of the same pixel values. I have found that this selection tool does not work very well on natural images, as there are not well-defined regions of the same pixel value, so I do not further discuss this tool. The button with the "A" adds text to an image. The next button is a magnification button, where you can change the scale at which you look at the image, without changing the image itself. Left clicks enlarge the image, and right clicks reduce the image. The final button is the dropper, which allows you to set the foreground color and background color for drawing operations by clicking on the current image. Left clicks set the foreground, and right clicks set the background.

Slide Shows

It is often nice to simulate some of the dynamics of the visual world. A slide show is a simple method of doing this that requires less technical sophistication and bandwidth than the methods discussed below. A *slide show*, in this context, is a sequence of still images that are presented to the participant at a relatively slow pace. Through a slide show, a sequence of events can be indicated and these sequences can carry a sense of change. In an excellent example, Loftus, Miller, and Burns (1978) used slide shows in early studies of eyewitness testimony.

Image sequences can be created in many ways on Web pages. The easiest way is to use the meta tags for redirecting a page to another page that can be placed in the head region of a Web page. The format for the meta tag is as follows:

<META HTTP-EQUIV="refresh" CONTENT="3; URL=nextpage.html">

This tag goes between the "<head>" and "</head>" tags on your Web page. The important elements of the code are the tags "refresh," "3," and "nextpage.html." The "refresh" tag indicates that the page will be refreshed or that a new page will be loaded without any action by the user. The "3" is the time in seconds that represents the delay before the page is refreshed. The "nextpage.html" is the URL, or address, of the page that will be loaded at the refresh time. The regular content of the page can then be anything that you would normally put on the page. To build a slide show, create a series of these pages, one for each image in your sequence, each refreshing to the next, with the last page not having the refresh tag. On each page, insert the image for that position in the sequence. On the last page, you can then add your questions using whatever method you want to collect your data.

Dynamic Media

VIDEO

Using dynamic media, such as video clips and animations, may be of great interest to psychologists in their research. The phenomena we study are dynamic, so it is sometimes good to have our stimuli match the phenomena.

General Video Issues

Until recently, the technology to reliably and quickly provide video formats through the Internet was not adequate for most researchers' tasks. Researchers and participants were plagued by large files, slow download speeds, intermittent stop and start of the video, and problems on the participant's end with video players and formats. Dictating the participants' ability to control the video is also of concern to the researcher. A particular research design may want the participant to be able to rewind and view the video again, whereas another research design may require the video to be watched at a particular rate and then not be accessible again by the participant. Additionally, researchers may not wish participants to be able to save the video on their local machine, and issues of copyright and stimulus security need to be considered. Now, however, many of the current formats for video delivery allow for this type of control, which was previously not available to researchers.

As briefly mentioned above, one issue that all researchers using a Web-based video format need to consider is the availability of the video player on participant's computers. Some video players integrate with the Web browser, and others are stand-alone programs. In either case, the video players need to be installed and available to the participant for such a study to be conducted.

File formats for video (e.g., AVI, MOV, MP4, MPG, and WMV) often produce large files. Common formats such as AVI and MPEG, which were originally developed for distribution via compact disc produce extremely large file sizes. For example, a 30-minute video clip captured with a high-definition camera and film (i.e., MiniDV tape) may easily take up 5 to 6 gigabytes of space on a hard drive, too large to be used on the Web. This clip must be converted, and in the process compressed, to produce a much smaller file. As is true with static images, compression will always reduce the quality of the videos. Because information is removed in the compression process, the resulting video may lose color quality, audio quality, frame rate playback, video playback size, or other features important to the researcher. You could edit the film clip into a much shorter version, and you could convert and compress it to a format appropriate for the Web. However, you will not have a video anywhere close to the original quality. Determining the quality needed is a task a researcher will need to engage in before simply placing a video online. If participants are expected to notice and discriminate between small details within the video, a large playback size and resulting video size will likely be needed. For studies in which the audio communications between those on the video are the key aspects, video playback size can be smaller without as much loss in audio quality.

An additional technological development that has allowed more Internet-based research is the increase in high-speed data connections across the United States. According to the U.S. Federal Communications Commission (2007), high-speed data access increased by 52% from June 2005 to June 2006, to a total of 64.6 million lines. This increase means that researchers, universities, and potential participants are more readily able to access video formats, whereas previously, the size limitation and the download speed would have made the routine use of Web-based video in research difficult. According to the same report, 99% of U.S. households have the potential to receive high-speed Internet access.

Determining the ability of a potential participant to reliably view and play the video is no longer as big an issue with the development of video players designed for Web-based delivery of video. However, for strict methodological purposes, a researcher will want to determine, and perhaps require, the use of a particular video player.

Streaming Versus Nonstreaming

One of the first approaches to overcoming the problems with video delivery was the development of streaming technologies. The ability to stream a video to the end user results in the user receiving the video in segments that are reassembled by the video player for play back and then usually discarded. Sometimes this may result in images appearing pixilated or as if they are constantly being reassembled. This is because the streaming video server and the client (i.e., the Web browser) are doing just that, assembling the video images as they are received and discarding viewed images. This allows for the video to begin playing while the remainder of the video is still being downloaded. This addresses many of the difficulties experienced by end users and alleviates the problem of having to wait for the entire file to download before beginning. Using streaming video allows the researcher to potentially control many of the aspects of the presentation, for example, varying the streaming rates on the basis of a participant's bandwidth. For researchers, this has the practical effect of shortening the amount of time the experiment takes from beginning to end and avoids periods of downtime when the participant is waiting for the stimuli to be delivered. However, although greater methodological control may be achieved, the end result could potentially be less consistency across participants. For example, when using a streaming method, the video player on the participant's machine must wait for the next segment to arrive before playing. Network congestion on the route from server to participant could potentially lead to the participant waiting several seconds while the video player waits for the next segment to arrive. In cases in which the video must be downloaded in its entirety before beginning, there will be no such stops and starts. Anyone who has watched streaming video from places such as cnn.com may have noticed this effect.

Additionally, streaming technology requires specific software hosted by the server used to conduct the study. The expertise needed to set up a streaming video server may be outside the realm of most psychologists, but the ability to convert video to a type appropriate for the Web is easily accomplished with today's commercial products. In addition to the cost of the server software, the streaming server and the resulting data storage needed require monitoring. Many commercial Web hosting companies allow for streaming video and provide these services.

The advantage of the nonstreaming approach for the typical researcher is the ability to film clips or scenarios themselves and load them on a Web site without having to deal with the added components, costs, or time required by using a streaming video server. However, the researcher will need to convert his or her video into a Web-based format. One of the most popular current formats is discussed below.

Current Internet-Based Format

The formats currently used by many Web sites for viewing video and animations provide a lot of the control needed by researchers to produce consistent methodologies across participants. Adobe's Flash Video format (FLV) is currently one of the best methods for delivering Web-based video. The Flash Video encoder allows a researcher to simply import the video and produce a FLV file. In the last couple of years, this particular video format, developed explicitly for delivery through the Internet, has resulted in a proliferation of Web-based video sites (e.g., youtube.com, blip.tv, videoegg.com)

ANIMATIONS AND OTHER DYNAMIC MEDIA

Flash

Many programs are currently available to produce video-like animations, allow for manipulations of objects by the participant, and dynamically branch or interact with the participant on the basis of participant responses. Adobe's Flash software, which originally was developed to produce animations for Web-based delivery and playback on Web pages, has recently begun to be used by researchers to provide both animation delivery and dynamic participant interactions (e.g., http://www.bbc.co.uk/science/humanbody/mind/, http://www.dectech.org/stian/learn1.html). The biggest advantage of using Flash as a method to develop and use dynamic media is the almost universal availability of the player required to view the Flash animations. The Flash Web player is routinely preinstalled with browsers such as Microsoft's Internet Explorer and Apple's Safari. Conducting experiments outside of a controlled lab setting removes one of the primary difficulties of retaining potential participants and reducing drop-out rates (see Figure 4.5).

Authorware

Adobe's Authorware program is another program that can be used to develop and deliver dynamic Web-based experiments (e.g., http://psychexps.olemiss.edu). However, unlike Flash, Authorware was not developed primarily as a Web animation program but as a program to develop presentation of materials that incorporate many interactive effects. For example, a participant may be asked to manipulate an object and move it to a location on screen or to sort and choose from various objects. Audio, video, and images may all be used along with many different response types (i.e., moving objects, clicking, check boxes, and text boxes) to produce interactive online studies. Many of these same results can be accomplished within Adobe Flash programming too, but with more of a learning curve. An advantage of using Adobe Flash is the cost: Adobe Authorware costs more that $500, even with an educational discount, whereas Flash is significantly cheaper.

FIGURE 4.5

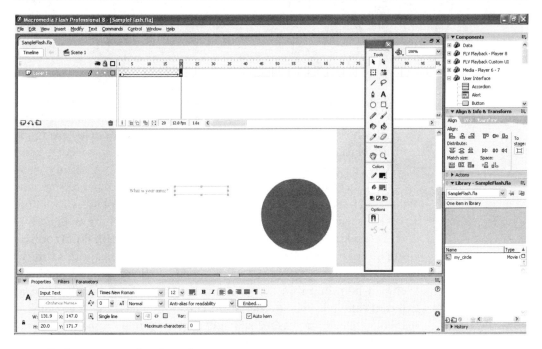

Flash programming interface.

CHOOSING AMONG METHODS

A researcher wishing to use video in a Web-based study will need to consider which details of the video presentation are important for the study. He or she will need to carefully decide between the following three factors: video playback size, audio quality, and video quality. The Adobe Flash video format appears to be the most widely distributable. However, new developments in technology mean that researchers conducting Web-based research will need to stay up-to-date on what is available.

The choice for other dynamic media is not as easily determined. Although Adobe's Flash program does provide both video and other interactive capabilities, other choices are available. Authorware may provide less of a learning curve than Flash, but the Authorware Web Player is not installed by default with Web browsers, requiring the participants to first download and install this program before a study can be run. Other ways of programming dynamic and interactive online studies also exist. Java can be used to create studies similar to those in Flash or Authorware (e.g., http://psych.hanover.edu/JavaTest/CLE/Cognition/cognition.html), but again there are considerations, such as the learning curve, consistent performance across operating systems, and new versions of Java.

Conclusion

This chapter has covered a variety of issues relating to the use of media on the Web in psychological research. Recommendations are scattered throughout the chapter: keep stimuli as simple as possible; use high-contrast, black-and-white images when possible; if color is needed, keep to as few colors as possible; and for video, consider the necessary playback size, image quality, and audio quality. Perhaps the main recommendation is to test images and video under Web-like circumstances, including across different machines, operating systems, and Web browsers. Conducting research on the Internet entails a loss of environmental control. Using Internet methods in a laboratory setting can allow for this control to be reestablished. Still, the loss of internal validity for studies on the Web can be minimized if stimuli are carefully designed; the gain in external validity can be tremendous. The tools to ease the development of media are ever increasing, so it is expected that the number of experiments using media should rapidly increase.

Additional Resources

Several links referred to in this chapter are available on the page for this chapter on the supplementary Web site (http://www.apa.org/books/resources/gosling). These links should help you use the examples in this chapter and gain access to the resources mentioned.

References

Federal Communications Commission. (2007). *Federal Communications Commission released data on high-speed services for Internet access* [press release]. Retrieved February 23, 2007, from http://hraunfoss.fcc.gov/edocs_public/attachmatch/DOC-270135A1.doc

Krantz, J. H. (2000). Tell me, what did you see? The stimulus on computers. *Behavior Research Methods, Instruments, & Computers, 32*, 221–229.

Krantz, J. H. (2001). Stimulus delivery on the Web: What can be presented when calibration isn't possible. In U.-D. Reips & M. Bosnjak

(Eds.), *Dimensions of Internet science* (pp. 113–130). Lengerich, Germany: Pabst Science.

Krantz, J. H. (2007). *Psychological research on the Internet.* Available from http://psych.hanover.edu/research/exponnet.html

Land, E. H. (1974). The retinex theory of colour vision. *Proceedings of the Roal Institute of Great Britain, 47,* 23–58.

Laugwitz, B. (2001). Web experiment on colour harmony principles applied to computer user interface design. In U.-D. Reips & M. Bosnjak (Eds.), *Dimensions of Internet science* (pp. 131–145). Lengerich, Germany: Pabst Science.

Levy, C. M. (1995). Mosaic and the information superhighway. A virtual tiger in your tank. *Behavior Research Methods, Instruments, & Computers, 27,* 187–192.

Loftus, E., Miller, D., and Burns, H. (1978). Semantic integration of verbal information into a visual memory. *Journal of Experimental Psychology: Human Learning and Memory, 4(1),* 19–31.

Rashband, W. (2003). *ImageJ.* Available from http://rsb.info.nih.gov/ij/index.html

Silverstein, L. D., & Merrifield, R. M. (1985). *The development and evaluation of color systems for airborne applications.* DOT/FAA Technical Report #DOT/FAA/PM-85-19. Washington, DC: DOT/FAA.

Wolfgang Neubarth

Drag & Drop

A Flexible Method for Moving Objects, Implementing Rankings, and a Wide Range of Other Applications

5

O nline environments can be programmed to make use of movable, drag & drop objects. A *drag & drop object* is a geometric figure (which may contain text, pictures, or both) whose position on the screen can be changed by the user. Typically, a user accomplishes this by (a) moving the cursor over the object with a mouse or other pointing device, (b) holding down button that "grabs" the object, (c) dragging the object to a new location by sliding the mouse or pointing device, and (d) dropping the object at the new location by releasing the button.

Drag & drop objects can be useful in several self-administered measurement domains of interest to social scientists; these include visual analogue scales (VASs; Couper, Tourangeau, & Conrad, 2006), ranking tasks (Neubarth, 2008; Thurstone, 1931), magnitude scaling (Lodge, 1981), ideal scaling (Barrett, 2005; Ryf, 2007), preference point maps (Harris Interactive, 2007), and various grouping or sorting tasks (Coxon, 1999). Even preference statements (Hensher, Rose, & Greene, 2005), choice analyses (Hensher et al., 2005), and the measurement of similarities or dissimilarities (Bijmolt & Wedel, 1995; Borg & Groenen, 2005) can benefit from using movable objects instead of simple hypertext markup language (HTML) forms.

This chapter explains how to implement a measurement framework for movable objects in Web questionnaires, how to define objects that can be dragged across all or predefined sections of the screen, and how to process the *x*- and *y*-coordinates of the objects. I provide code that restricts the range of the drag & drop objects to a predefined area. I also provide a function that allows one to align the objects smoothly after they are dropped. Finally, I suggest some practical applications for the movable object method and provide some further resources.

This chapter will be useful to researchers who want to use drag & drop functionality in their Web research. It will also be useful for those who want to learn more about the technical basics of the document object model (DOM; see Keith, 2005, and Marini, 2002). The DOM allows objects to be moved and their positions to be saved into a data set. The chapter will also be useful to those who want to learn about the research possibilities afforded by drag & drop functionality.

Drag & drop will allow you to set up objects on an HTML page that can then be moved by the participants. The information about the movement process or the end state can be saved and made available for data analysis. This method works with simple HTML and JavaScript (Flanagan, 1997), so it enables drag & drop features to be added to a simple Web questionnaire. In contrast to other methods, like Flash, Java applets, or scalable vector graphics, no additional plugins for the browser are needed. Many complicated tasks, which were not previously possible in self-administered interviews, become feasible with this method of data collection (e.g., magnitude scaling, ranking tasks, and every kind of sorting or ordering task).

In cases in which a researcher could have participants use either numbers to represent rankings or drag & drop technology, the latter is often preferable because it is more intuitive to participants and is less subject to error. When participants must type in a number or click a radio button adjacent to a number to represent their rankings, some may become confused about whether, for instance, the number "1" represents a high ranking or a low ranking. With drag & drop technology and clear instructions, participants easily understand that higher ranked items are to be dragged to the top of the scale. For sorting tasks that use cards in a real-world, non-virtual environment (e.g., the California Adult Q-set; Letzring, Block, & Funder, 2004), asking participants to sort statements by selecting numbers would be cognitively overwhelming, whereas asking them to sort statements with drag & drop would be as easy as sorting cards in real life.

From a technical point of view, the implementation of movable objects and their measurement is not very complicated, although some code is needed. The functions described here are part of the DOM (Keith, 2005; Marini, 2002), which is in standard JavaScript, making it widely useable on virtually all common browsers.

One disadvantage of the DOM is that the interactive Web content is not accessible by screen readers, making it unavailable to people who are blind; however, the Web Accessibility Initiative—Accessible Rich

Internet Applications Initiative (Schwerdtfeger & Gunderson, 2006) is currently working to solve this problem.

A second drawback of this method is that the interfaces researchers create with the program code provided here are not part of the HTML standard form definition. So the participants need to learn new functionalities' answering formats, which might affect the validity of the collected data sets.

Teaching Example

To help readers understand how to create drag & drop functionality, I provide the code to implement a simple design with two movable rectangles (see Figure 5.1). For the downloadable script code, please see the

FIGURE 5.1

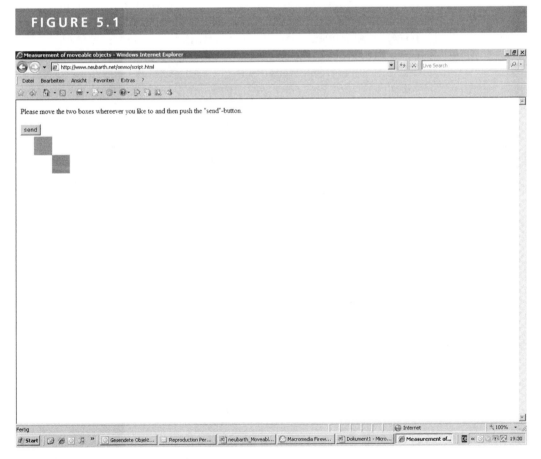

Browser view of the teaching example.

supplementary Web site for this chapter. When pushing the "Send" button, the *x*- and *y*-coordinates of both rectangles will be submitted by the method "get," so they can be seen in the address line of the browser.[1] Please feel free to test the drag & drop examples provided in this chapter firsthand by visiting the supplementary site for this book. You may also download the source code (usually by clicking Data -> Save as . . .) provided on the supplementary page to change and experiment with the examples for your own purposes.

To teach readers about drag & drop functionality, I next describe a simple script that generates a Web page with two rectangles that can be moved to any location on the screen. Figure 5.1 shows what the Web page looks like before a user drags the rectangles to different locations. When the user is finished moving the rectangles and clicks a "Send" button, the script reports the *x*- and *y*-coordinates of both rectangles.

The code for the script is divided into three major segments, which are presented in Figures 5.2, 5.3, and 5.4. (A fourth segment, presented in Figure 5.5, handles some fine tuning of object movement and is discussed in the Reordering the Objects section.) The script contains standard HTML code; readers unfamiliar with HTML can refer to Fraley (2004). The initial properties of the two rectangular objects are defined with Cascading Style Sheets (CSS) between the "<STYLE>" and "</STYLE>" tags. The JavaScript between the "<SCRIPT>" and "</SCRIPT>" tags provides the functions for moving the objects and keeping track of their *x*- and *y*-coordinates. Readers unfamiliar with CSS and JavaScript can refer to Keith (2005) and Marini (2002; see also the additional resources at the end of this chapter for Web-based tutorials).

In the first part of Figure 5.2 (Lines 7–22), the objects are initialized, and static properties of the objects are defined. When computation is needed to set the objects because, for example, they should be centered at the beginning, properties are defined in the "init" function (Lines 17–24).

The initialization process is differently implemented in common browsers. The statement "var div2= eval(document.getElementById ("div2"));" is needed when using a Mozilla derivative but is obsolete for Internet Explorer. Mozilla also requires setting string values ending with a unit of measurement for each object property, which is "px" in the example shown here. The code in Lines 23 and 24 captures all mouse clicks within the browser window. All properties of the mouse click are also stored in a corresponding JavaScript object.

---

[1]For our teaching example, we want the coordinates to appear directly in the address line of the browser. This simplifies the understanding and testing of the functionalities of the script code. Method "post" is more secure and therefore the preferred method in a real survey.

FIGURE 5.2

```
3    <html>
4    <head>
5    <title>Measurement of moveable objects</title>
6    <STYLE>
7    #div1, #div2
8      {
9       height:40px; width:40px; position:absolute; z-index:1;
10     }
11   </STYLE>
12   <SCRIPT>
13   var zIndexTop = 10;
14   var activex;
15   var activey;
16   function init() {
17   var div1= eval(document.getElementById("div1"));
18   div1.style.left = 40 +"px";
19   div1.style.top= 80 +"px";
20   var div2= eval(document.getElementById("div2"));
21   div2.style.left = 80 +"px";
22   div2.style.top= 120 +"px";
23   document.onmousedown = startDrag;
24   document.onmouseup = stopDrag;
25   }
```

Part 1 of the teaching example.

The second part of the example, shown in Figure 5.3, handles the actual movement of the objects in three functions: "startDrag," "drag," and "stopDrag." These functions are the core part of the movable object script and would not need to be changed if additional objects were added.

The "startDrag" function (Line 26) first checks whether a valid object is clicked,[2] then sets it on top over all other objects by incrementing "zIndex" and stores the "old" positions of both click and object. The "drag" function (Line 41) computes the difference between the old and the actual mouse position and adjusts the position of the object to it. The "parseInt" function (Line 46) is necessary because Mozilla stores the values as string variables. So to compute with these measures, they need to be transformed into integers and back every time.

---

[2]Objects that should not be movable but have an ID need to be disabled here. In this example, we do not want to make the text movable.

FIGURE 5.3

```
26  function startDrag (e) {
27  if(!document.all) { objekt=e.target.id; }
28  else {  objekt=event.srcElement.id; }
29  if (objekt =="text" )
30  {objekt=false}
31      if(objekt) {
32          if(!document.all) {event=e;}
33          zIndexTop++;
34          document.getElementById(objekt).style.zIndex=zIndexTop;
35          startX=event.clientX;
36          startY=event.clientY;
37          XpositionAlt=document.getElementById(objekt).style.left;
38          YpositionAlt=document.getElementById(objekt).style.top;
39          document.onmousemove=drag;
40          return false; }}
41  function drag (e) {
42      if(objekt) {
43  if(!document.all) {event=e;}
44  Xvalue=event.clientX-startX;
45  Yvalue=event.clientY-startY;
46  //if (parseInt(XpositionAlt)+Xvalue < 400)
47  document.getElementById(objekt).style.left=parseInt(XpositionAlt)+Xvalue + "px";
48  document.getElementById(objekt).style.top=parseInt(YpositionAlt)+Yvalue + "px";
49  }
50  return false;}
51  function stopDrag () {
52          if (objekt) {
53          document.move[objekt +"xcor"].value =parseInt(document.getElementById(objekt).style.top);
54          document.move[objekt +"ycor"].value =parseInt(document.getElementById(objekt).style.left);
55          //reorder()
56          objekt=false;       }
57          return false;     }
```

Part 2 of the teaching example.

The "stopDrag" function (Line 51) sets the hidden form fields, defined in Part 3 of the code (see Figure 5.4, Lines 85–89), to the end state of the moved object. (For a description of hidden form fields, see Fraley, 2004, chap. 4.) Then "stopDrag" releases this object again, so it is not moved any further until it is clicked again.

The third and last part of the example, shown in Figure 5.4, contains the classic HTML portion of the script. It begins (Line 78) by loading the "init" function from the JavaScript, which displays the rectangles in their original position, defines the object IDs with "<div>" tags, and finally the HTML form, which uses the hidden fields, to submit x- and y-coordinates of the movable objects. For the purpose of this teaching example, the method attribute "get" is used to display the submitted coordinates in the address field of the browser to show what these numbers look like. In a real application of this script, the method "post" instead of "get" would be used to send the coordinates to a data file.

FIGURE 5.4

```
76   </SCRIPT>
77   </head>
78   <body onLoad="init();">
79   <div id="div1" style="background-color:#FF0000;">  </div>
80   <div id="div2" style="background-color:#0099FF;">  </div>
81
82   <div id="text">Please move the two boxes whereever you like to
83   and then push the &quot;send&quot;-button.</div>
84   <form name="move" method="get">
85   <input type="hidden" name="div1xcor">
86   <input type="hidden" name="div1ycor">
87   <input type="hidden" name="div2xcor">
88   <input type="hidden" name="div2ycor">
89   <input type="submit" value="send">
90   </form>
91   </body>
92   </html>
93
```

Part 3 of the teaching example.

FIGURE 5.5

```
58   function reorder() {
59   element = eval(document.getElementById(objekt));
60   window.clearInterval(activex);
61   activex = window.setInterval("movex()",1);
62   activey = window.setInterval("movey()",1);
63   }
64   function movex(){
65   if (parseInt(element.style.left) % 50 != 0)
66   element.style.left =  parseInt(element.style.left) + 1 + "px";
67   else
68   window.clearInterval(activex);
69   }
70   function movey(){
71   if (parseInt(element.style.top) % 50 != 0)
72   element.style.top =  parseInt(element.style.top) + 1 + "px";
73   else
74   window.clearInterval(activey);
75   }
```

Extension of the teaching example.

The HTML portion of the script would be executed, even if the user has disabled JavaScript or is using a very old browser. Part 3 of this basic example completes the functional script code.

RESTRICTING THE RANGE

In VASs, only movements on the *x*-axis are needed. Also, only a certain range on this axis should be allowed to place the slider. To prohibit movements on the *y*-axis, Line 48 (see Figure 5.3) just needs to be commented out by two slashes ("//"). By adding

$$\text{if} \left(\text{parseInt} \left(\text{XpositionAlt} \right) + \text{Xvalue} < 400 \right)$$

into Line 46, the maximum left position for the rectangles is 400 pixels on the screen. The same *if*-clause restrictions for the left side of a VAS are easy to implement.

REORDERING THE OBJECTS

After a participant releases an object, it is sometimes useful to rearrange the objects a bit. For example, they can be adjusted on a grid or rearranged to prevent overlapping. This can be done by setting the "document.get ElementById(objekt).style.top" variable to the needed value for the *y*-axis, but this method is rather crude. Participants like a smooth rearrangement better than an absolute setting. This can be accomplished by activating the "reorder" function in Line 55 of the section of the code shown in Figure 5.3. The two slashes ("//") in Line 55 need to be deleted to implement the "reorder" function. The code provided in Figure 5.5 needs to be added to the script as well. Now the rectangles are rearranged smoothly to a grid with 50×50 pixel each time they are released. I offer this example of the "reorder" function just to point out its basic functionality; this function has many more helpful uses.

EXTENDED STORAGE OF PARADATA

The example provided in this chapter generates only a small amount of data. Usually data sets gained by this procedure are massive, not quadratic, in shape and need simplification. Therefore, data can be resolved during data collection or while data clearing. In the code example shown in Figures 5.2, 5.3, 5.4, and 5.5, the simplification is already handled in the data collection phase. Only the *x*- and *y*-coordinates of the objects in their end state are considered. In the code of Figure 5.3, all former coordinates are overwritten by the last valid entry.

Of course, it is also possible not to overwrite former coordinates but to append the new data points marked by a separator. Then all *x*- and *y*-coordinates, the object ID, plus the corresponding time stamps (measured in milliseconds) would be obtained.

Data analysis for the code shown in Figures 5.2, 5.3, and 5.4 is fairly simple because only the end states of the objects are submitted and hence can be analyzed. Even in this case, it might be appropriate to define sections of the page to classify the end states of the objects to certain categories. When analyzing a preference map, one will find it is helpful to group the *x*- and *y*-coordinates to substantive areas on the map. For a ranking task, all the coordinates need to be recoded to the corresponding rank.

For the analyses of complex data, which contain the complete history of clicks done by a participant, it is necessary to parse the data before interpretation and analysis. Appropriate aggregation levels need to be defined before analyzing the data. The researcher might like to group coordinates to certain areas and count the number of clicks into this region. Respondents' number of position changes during the response process can be saved as a variable for each object, too. This measures users' certainty about the positions of certain objects. The response time used for each object can also result in a variable for each object. It is calculated by adding the difference between two clicks to the amount of time for the clicked object. The number of variables obtained for just one object depends on the level of accuracy needed for each project. A complex analysis might require information on five variables of paradata (the *x*- and *y*-coordinates, the number of clicks on the object, the response time used for the object, and the distance the object was moved during the answering process), but a simpler analysis (e.g., from a ranking task) would result in just one variable.

Practical Applications

In the preceding section, I presented a teaching example without real practical use to demonstrate the basic functionalities of the movable object method. Using two examples, I now demonstrate how this technique can be applied in research. Figure 5.6 shows a ranking application implemented with the movable object method, and Figure 5.7 shows an application to scaling.

Usually, ranking measures are complex, expensive, or difficult to obtain in self-administered settings. For example, Rokeach (1968) sent out his value survey, containing core values of the American society, with 18 gummed labels on one side of a paper questionnaire. On the other side of the sheet, 18 predefined boxes were printed to indicate the rank. To indicate the order of their personal values, participants were required to paste the gummed labels onto the predefined boxes. Most other ranking

FIGURE 5.6

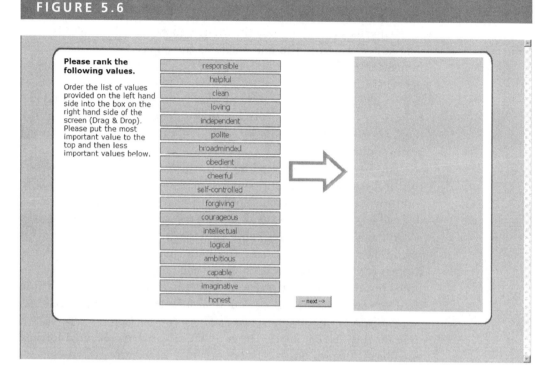

Ranking application with the use of movable objects.

and sorting methods are equally complex. However, with the movable object method, rank orderings are easy to drag & drop for the respondents, even for large sets without a personal interviewer. This ranking application, using movable objects, is available from, for example, Globalpark.[3] A task as complex as that illustrated in Figure 5.7 would be hard to implement in any traditional questionnaire format but is relatively straightforward using drag & drop technologies.

At the beginning of the task, all objects, which might be logos of car producers, different travel destinations, or the Rokeach Value Scale for example, are arranged on the left side of the screen. When a participant drags one of the objects, an arrow appears and points to the scale on the right-side. This allows researchers to obtain metric information for all objects on one scale. For the analysis, only the measure of the arrows on the scale are relevant. The second dimension has no use for data

---

[3]Globalpark allows universities, research departments and colleges further education reduced license rates (http://www.unipark.info/1-1-home.htm). Commercial customers need to apply for a usual license (http://www.globalpark.co.uk/).

FIGURE 5.7

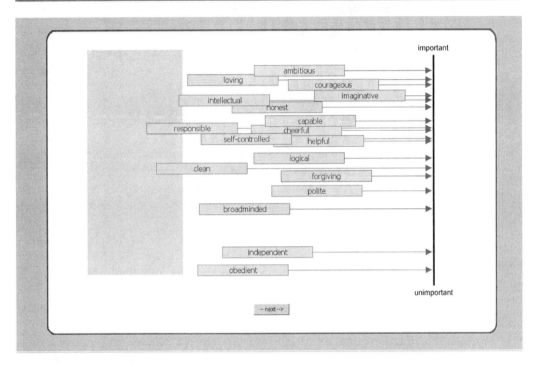

Ideal scaling application with the use of movable objects.

analysis, but it allows the participants to state tied ranks for the objects. A sample script for the Ideal Scale is available at the supplementary Web site for this chapter.

The applications presented here are just two examples of the many possible uses of drag & drop technology. Finally, the implementation of drag & drop technologies into Web-based questionnaires enables a new toolbox of data collection for behavioral research.

Additional Resources

- DOM tutorial at http://www.w3schools.com/htmldom/dom_intro.asp
- Articles and tutorials about DOM at http://xml.com/pub/rg/DOM_Tutorials
- Various control panels at http://www.blueshoes.org/en/javascript/

References

Barrett, P. (2005). Person–target profiling. In A. Beauducel, B. Biehl, M. Bosnjak, W. Conrad, G. Schönberger, & D. Wagener (Eds.), *Multivariate research strategies: Festschrift in honor of Werner W. Wittman* (pp. 63–118). Aachen, Germany: Shaker.

Bijmolt, T. H. A., & Wedel, M. (1995). The effects of alternative methods of collecting similarity data for multidimensional scaling. *International Journal of Research in Marketing, 12,* 363–371.

Borg, I., & Groenen, P. J. F. (2005). *Modern multidimensional scaling theory and applications* (2nd ed.). New York: Springer.

Couper, M. P., Tourangeau, R., & Conrad, F. G. (2006). Evaluation the effectiveness of visual analog scales—A Web experiment. *Social Science Computer Review, 24,* 227–245.

Coxon, A. P. M. (1999). *Sorting data: Collection and analysis.* Newbury Park, CA: Sage.

Flanagan, D. (1997). *JavaScript: The definitive guide* (2nd ed.). Cambridge, England: O'Reilly.

Fraley, R. C. (2004). *How to conduct behavioral research over the Internet: A beginner's guide to HTML and CGI/Perl.* New York: Guilford Press.

Harris Interactive. (2007). *Online survey enhancement concept testing.* Retrieved May 20, 2007, from http://hpolsurveys.com/enhance.htm

Hensher, D. A., Rose, J. M., & Greene, W. H. (2005). *Applied choice analysis: A primer.* Cambridge, England: Cambridge University.

Keith, J. (2005). *DOM scripting: Web design with JavaScript and the document object model.* Berkeley, CA: Apress.

Letzring, T. D., Block, J., & Funder, D. C. (2005). Ego-control and ego-resiliency: Generalization of self-report scales based on personality descriptions from acquaintances, clinicians, and the self. *Journal of Research in Personality, 39,* 395–422.

Lodge, M. (1981). *Magnitude scaling.* Newbury Park, CA: Sage.

Marini, J. (2002). *Document object model: Processing structured documents.* Berkeley, CA: McGraw-Hill.

Neubarth, W. (2008). *Präferenzdaten online* [Preference data online]. Cologne, France: Halem.

Rokeach, M. (1968). A theory of organization and change within value–attitude systems. *Journal of Social Issues, 24,* 13–33.

Ryf, S. (2007). *Multidimensionale Skalierung in der Marktforschung: Möglichkeiten und Grenzen* [Multidimensional scaling in market research: Opportunities and limitations]. Zurich, Switzerland: Dissertation at the Psychological Institute of the University of Zurich.

Schwerdtfeger, R., & Gunderson, J. (2006). *Roadmap for accessible rich Internet applications (WAI–ARIA Roadmap).* Retrieved May 20, 2007, from http://www.w3.org/TR/aria-roadmap/

Thurstone, L. L. (1931). Rank order as a psychophysical method. *Journal of Experimental Psychology, 14,* 187–201.

STUDYING INTERNET BEHAVIOR

Elizabeth Mazur

Collecting Data From Social Networking Web Sites and Blogs 6

This chapter describes the use of social networking Web sites and blogs—two types of social media that can be used productively in behavioral research on the Internet. *Social networking sites* are online venues where members can create and post content to profiles (i.e., lists of demographic information and personal interests constructed by completing forms within the site) and can form personal networks that connect them to others using tools embedded in the social software. *Blogs,* published either on social networking sites or on separate public hosting Web sites, are reverse-chronological, time-stamped, online journals on a single Web page. More text oriented than social profiles, blogs, like social network pages, are often interconnected; many writers read and link to other blogs, referring to them in their own entries. Because of this trend, the interconnected blogs often become part of social communities with their own culture, often referred to in *toto* as the *blogosphere.* The more comprehensive term *social media* refers to Internet services that facilitate social human contact. Besides blogs and social networking sites, social media include the creation and posting of digital material (e.g., photographs, music, videos) on specialized Web sites, personal Web pages, instant messages, online discussion groups, wikis, and virtual worlds. Because of my research experience, potential information overlap with other types of social media, and online social networks' relatively text-based nature, this chapter focuses on social networks and blogs as sources of data. However, I discuss other types of social media when the issues are relevant or the examples useful. Specifically, this chapter describes searching for, and sampling from, social network profiles and blogs, human (rather than computer) content coding and analysis, and intercoder reliability. Although researchers may directly contact users of social media, this chapter focuses on noninterfering measures; the topic of direct interaction with users can be found in Johnson (chap. 10, this volume). Although this type of research is unobtrusive, ethical issues are still important (see chap. 16, this volume). In sum, this chapter will be most useful to behavioral scientists who wish to use social networking sites and blogs to unobtrusively study text, music, and image-based data on the Internet. Because content analysis is used extensively in social media research, familiarity with the technique will be helpful.

Analysis of social networks and blogs, as well as other social media, will allow you to systematically and objectively analyze textual, visual, and oral content created by large and diverse populations. Researchers can study individuals and groups within a naturalistic setting without the presence of an intrusive researcher. Both quantitative and qualitative analyses are possible because social networkers and bloggers create narratives that often disclose information about themselves that offline would normally be part of a slower and more private process of acquaintanceship but online are posted frequently and often in full public view.

Besides the writing of narratives and the sending and receiving of messages, social networking sites and blogs encourage *friending* (connecting with others online, often through network and profile-surfing) and the posting of photos, music, and videos; thus, they can be mined for data on oral and visual culture, presentation of self and identity, language use, and interactive social dynamics, among other topics. Analyses can help answer questions derived initially from observations of offline behavior, as Huffaker and Calvert (2005) and Thiel (2005) demonstrated in their studies of gender influences on identity construction and language use in instant messaging and teenagers' blogs, respectively, and as Richards and Mazur (2008) showed in their comparison of characteristics of adolescents' offline and online friendships. Alternatively, researchers can study the specific phenomenon of content creation on the Web, as did Stern (2002) and Mazur and Kozarian (in press) in their analyses of girls' homes pages and young adults' blogs, respectively, as performance and as projections of identity and image.

Considered a force in media, entertainment, politics, advertising, and marketing (McCoy, 2008; Stelter, 2008b), online social networks, in particular, are ripe for the study of technology-integrated communication. Thus, researchers may want to directly study processes of communication on the Internet, as did Rodham, Gavin, and Miles (2007) in their study of interaction among nonprofessionals on a self-harm message board.

The content of blogs tends to be more narrative driven than that of social networks and other social media, more similar, ironically, to private journals, and thus particularly amenable to traditional content analysis. Most bloggers write about their life and experiences, but you can find blogs that focus on almost any cause or passion, such as politics, entertainment, business, physical and mental health and illness, family dynamics, and religion. Due to its size and occasional ambiguity about which blogs remain active, the blogosphere is usually not studied as a whole, but rather with respect to certain topics or blog clusters, which may reflect social networks of professional affiliations, common interests, or personal friendships (Schmidt, 2007). Thus, you can analyze blogs written by persons of a certain age or who share other demographic commonalities, or you can examine blogs by topic, as did Singer (2005) in her study of blogs concerning government and civic affairs.

You can easily collect data about individuals' or groups' perceptions of naturally occurring historical events, such as wars, elections, or public figures. Certainly, blogging about political topics appears to be a particularly popular topic of analysis (Bahnisch, 2006; Singer, 2005), followed by so-called "hipster" lifestyle blogs, tech blogs, and blogs authored by women (comScore Media Metrix, 2005). You need not study only contemporaneous events, because archives of blogs and posts from newsgroups and social networks (albeit dating back at the earliest to 1995) are available and searchable; Google Groups (http://groups.google.com/) is one easily accessible source. The potential of social media as historical archives is only going to grow.

With some exceptions (e.g., the very poor and the very old), access to a diverse sample is more likely to be the rule than the exception. Although adolescents and young adults are the most avid users of social networking Web sites, their popularity is gaining with adults. The fastest growing demographic group of social networkers is now over 25 years old, with persons in that age group composing 68% and 71% of MySpace, and Friendster users, respectively (comScore Media Metrix, 2006; Skiba, 2007). In fact, Internet users between the ages of 35 and 54 account for almost 41% of the MySpace visitor base, although age demographics do vary by hosting Web site (comScore Media Metrix, 2006). Of American online teens, 55% have posted profiles to at least one social networking site (Lenhart & Madden, 2007; Lenhart, Madden, Macgill, & Smith, 2007), and they are generally a diverse group, although somewhat skewed toward females (58%). Fifty-five percent report household income less than $50,000, and 58% consider themselves to be a race other than White (Lenhart & Madden, 2007). Although comparable social network posting data for American adults is unavailable, the percentage that has placed a profile on at least one social networking site may or may not be similar to the 29% of Canadian adults who have done so (Mindlin, 2008; Surridge, 2007).

Even more people visit than post on social networking profiles, and this popularity can lead to potentially productive lines of research about the messages that are being conveyed to and by users, such as representations of mental illness (Wollheim, 2007). Twenty-four percent of American online adults have reported visiting a social network within the past 30 days, as have 20% of online adults worldwide (Havenstein, 2007). Notwithstanding language barriers, this international diversity potentially lends itself well to comparative international studies of, for example, differences in self-presentation or the management of social relations.

For text-based researchers, blogs (estimated at 70 million in number) typically present richer data than do social profiles, but they appear to be less popular than personal pages on social networks. Only about 55% are active and updated at least once every 3 months, although the blog tracking firm Technorati estimates that 120,000 new ones are created every day (Sifrey, 2007). Thus, the numbers are still impressive for data sampling; about 28% of online American adolescents and 8% of online American adults have created a personal blog, and 49% and 38% of online adolescents and adults, respectively, read them (Lenhart & Fox, 2006; Lenhart et al., 2007). The population of bloggers is relatively diverse in that more than half of American bloggers are under age 30 and about 50% are female, and they are less likely to be White than is the general American population (Lenhart & Fox, 2006). And blogs, like social networks, can gain researchers access to international populations. In fact, 37% of all posts are written in Japanese, followed closely by English at 36% (Sifrey, 2007), with Chinese (8%), Italian (3%), and Spanish (3%) composing the next most common posting languages. Thus, as with social networking sites, notwithstanding potential language obstacles, the international ubiquity of blogs gives researchers the ability to instigate comparative research designs.

A serious drawback is that researchers of social media need to mistrust, to some extent, the credibility of their informants. Because users have personal reasons for interacting with social media, social networkers and bloggers may be more likely than active research participants to misrepresent personal information, especially their age, if they post it all. This is especially likely for children and adolescents. For example, MySpace and Facebook do not permit users under ages 14 and 13, respectively, and MySpace automatically classifies as private the profiles of members ages 14 to 17 (Stelter, 2008a). However, because Webmasters presently have no reliable way of validating a user's age, younger children may still be posting, and users will likely be able to circumvent some of the MySpace changes. MySpace adults also may alter their ages to ones they perceive as more socially desirable, for example, if they are searching for romantic connections or new employment. Even for honest users, most social networking sites, such as MySpace and Facebook, allow users to not reveal their birthday, and the default settings typically limit the demographic information displayed in a person's profile. Although much of the information found in a person's social profile is likely to be honest and unedited, social networkers and bloggers of any age may falsify other personal data or embellish or delete information in their narratives. For example, Oblinger and Hawkins (2006) interviewed a college student who cited a friend's Facebook portrayal as dishonestly presenting a "party animal" rather than the more truthful "nerdy wallflower" (p. 14).

Less frequently, users may fabricate the content of another person's social profile or blog without their consent or may manufacture a completely false one. The *New York Times* has reported that many profiles on MySpace do not represent actual people and that "friending" from attractive women may conceal advertisements for pornographic Web sites (Stone, 2007). Two recently publicized examples of profile fabrication include an Internet prankster who was able to post a sham Facebook profile of Bilawal Bhutto Zardari, the son of former Pakistan Prime Minister Benazir Bhutto. This event occurred despite Facebook's claims of examining multiple criteria to determine whether a profile is authentic (Nizza, 2008). Multiple persons are reported to have created a fake identity for a fictitious teenage boy who harassed a young female adolescent on her MySpace page (Cathcart, 2008). These are unlikely to be isolated events.

Implementing a Social Network or Blog Analysis

Readers of this chapter will have varying levels of comfort and experience with using social media and with the logistics of online data collection. But implementing a social network or blog analysis will be as easy or as challenging as you determine it to be, based on the scope, complexity, and conceptualization of the research questions. One of the initial decisions will be to select appropriate social profiles and blogs to sample.

SEARCHING FOR SOCIAL NETWORKING PROFILES AND BLOGS

One of the most direct, but not necessarily the most comprehensive, ways to locate social networking profiles and blogs pertinent to your research topic is to search directly on those sites reported to be most popular with the population of interest. At the time of this writing, the three largest social networking sites were reported to be MySpace (118 million active members), Facebook (161 million members), and Bebo, the most popular site in Europe with 40 million active users (Ante, 2008; McCoy, 2008; Shrivastava, 2007; Stelter, 2008c). As of 2005, the most recent data available, the five most frequently visited blog hosting sites were Blogspot, Livejournal, TypePad, MySpace, and Xanga, each with more than 5 million visitors per quarter year (comScore Media Metrix, 2005). These user numbers suggest that these sites are likely to include many social profiles and blogs that fit your specific demographic criteria. Livejournal, for example, contains many subcultural groups ranging from persons with disabilities (Goggin & Noonan, 2006) to Goths from the United Kingdom (Hodkinson, 2006).

However, depending on your population of interest, it may be more productive to search relatively smaller social networks, some of which host blogs, with reputations for distinctive cultures. For example, if you want to include social profiles of Asian-Pacific young adults in your analysis, current information suggests that Friendster, Southeast Asia's top social networking site for both adolescents and adults, would be at least one site from which you would want to sample (McCoy, 2008). Other specialized and fast-growing social networking sites include Black-Planet (for African Americans), Care2 (for social activists), DeadJournal ("the journals that nobody else wants to see, or even host"), Eons (ages 50-plus), Faith Base (Christian), GLEE (gay, lesbian, and bisexual communities), Habbo (a virtual world "hangout for teens"), Hi5 (popular with Spanish speakers), LinkedIn (for making business contacts), Ning (for creating and customizing one's own social network), TG Times (for transgender people), and YouTube (video sharing).

To locate these more specialized hosting Web sites, print and online newspapers often publish articles about rising hosting Web sites and changes in their user base (e.g., McCoy, 2008). Easily accessible and relatively frequent Internet reports from organizations such as comScore (http://www.comScore.com), the Pew Internet & American Life Project (http://www.pewinternet.org), Nielsen/NetRatings (http://www.nielsen-netratings.com/), and Technorati (http://technorati.com) can often point you to the most popular social network and blogging Web sites for persons of different ages and interests. Professional organizations may help one locate relevant social networking sites and blogs; Singer (2005) chose journalism blogs from a list maintained by the American Press Institute published on http://www.cyberjournalist.net.

Besides visiting hosting Web sites already known to be populated by a particular sample, such as the social networking Web sites described above, you can also often unearth pertinent content through regular Internet search engines. For example, Calasanti (2007) located 96 anti-aging Web sites by simply typing the key phrase "anti-aging" into the AltaVista search engine (http://www.altavista.com). However, as she acknowledges, this method tends to be inefficient, with many inactive and irrelevant sites listed, although such a search may help establish which hosts are popular among the population of research interest. For example, my 2008 Google search of "democratic primary" + "social network" was somewhat successful in that among the first 10 results, there were listed 5 potentially relevant social networking profiles located on the blog sites Ning and MyPageHost.net thus suggesting I proceed to those sites to locate political blogs on this topic. However, the success of a word or term search varies with the topic. For example, among the top 10 results of a Google search of "teen + blog" were 1 porn site (listed first despite the computer's "moderately safe search" setting), 2 blogs written by teenagers, 5 library Web sites oriented to teens, 1 Web site of a popular teen magazine, and 1 advertisement for a traffic school.

Another search option for locating blogs in particular, as well as readers' comments to them, are specialized engines and directories such as http://blogsearch.google.com (allows you to choose a publication date; e.g., last hour, last 12 hr, last day, past week, past month, or anytime or choice of dates), http://www.blogsearchengine.com (focuses on "high-quality" blogs presented in categories), http://portal.eatonweb.com (presents blogs in the categories "fastest gainers," "fastest sinkers," and "newest additions"), http://www.thebobs.com (features an international encyclopedia of blogs to browse through by country, language, or subject matter), http://www.blogpulse.com (enables you to search by topic of interest and to discover links among blog entries), and http://technorati.com/ (has the tagline "What's percolating in blogs now"). Ali-Hasan and Adamic (2007), for example, used the last two sites for their application of social network analysis to the blogosphere.

Once a social profile or blog is located, the reader may or may not need to log in. If you want to directly search social networking profiles or blogs from within hosting Web sites, however, most require readers to open an account with a user ID and a password. Even for those that allow access and search without login, such as the blog hosting site GreatestJournal, maintaining a user account, which is usually free, typically renders the site easier to navigate. Many sites require the login ID to be an e-mail address. I suggest opening a free account on a public e-mail site, such as Yahoo! or Gmail, solely for the research project. For ease of use by principal investigators and coders, use the same account and password for all the blogging and social network sites one is to study. You need not complete a profile or write a blog to maintain an account.

Once on the hosting Web site, you can typically search through user profiles by age, sex, location, title or topic, date of first entry, date of last update, interests, school attended, and photos, although sites vary in these choices and their organization. Some sites, such as YouTube, are searchable by content; for example, Wollheim (2007) accessed video portrayals and discussions of mental illness by searching within the site by specialized categories such as "schizophrenia," "eating disorders," and "suicide." With the exception of Technorati, these search engines operate at the level of text found in blog postings. However, most blog posts also contain *tags,* user-generated labels that categorize and retrieve Internet content. At present, there are two major ways to locate blog information classified through tags. Technorati scans the blogosphere for tags that appear in blogs postings, which are created by the person who created the post. Over one third (35%) of all posts tracked by Technorati use creator-assigned tags (Sifrey, 2007). Google also features tagged social media in its results pages.

How easy it is for the researcher to learn the demographic characteristics of the blogger or social network user depends on the particular hosting Web site and the individual poster. Besides the initial basic search categories, a noncomprehensive list of the information that many user profiles display includes author's age, gender, marital status, sexual orientation, religion, location, ethnicity, number of children, education and schools attended, occupation, income, interests, and body type. For example, if I were to sample the blogs of single fathers, I could search within a hosting Web site by requesting only blogs that state "male" as "gender," "single" as "marital status," and at least "1" for "number of children." However, be aware that only some or none of this information may be available to the researcher because bloggers and social network users may set part or all of their profiles to private.

As mentioned earlier, one reason that blogs and social networks are particularly interesting to study in terms of social communication is that they allow researchers to know who is commenting on a writer's post and what other blogs or social profiles the initial writer finds relevant. Blogs and social network pages typically connect to others through links usually, although not always, called "friends," "favorites," or "comments." Thus, once you have located one blog or social profile useful for your research, it is often easier to find others. These connections are made possible through trackbacks and pingbacks. As defined by Dieu and Stevens (2007), a *trackback* is an acknowledgment that is sent via a network signaling ping from the originating site to the receiving site. The receptor often publishes a link back to the originator indicating its worthiness and may display summaries of, and links to, all the commenting entries below the original entry. Conversations spanning several blogs can then be followed. A *pingback* is a signal sent from one site to another that allows a blogger to notify another blog of its latest entry by posting its permalink directly in the content of the new blog entry. It is similar to a link in that when the second site receives the notification signal, it automatically checks the initial sending for the existence of a live incoming link. If that link exists, the pingback is recorded successfully (Dieu & Stevens, 2007).

SAMPLING

For a blog or social network analysis to be generalizable to some population of persons, the sample for the analysis should be randomly selected. This generally requires a two-step process: (a) sampling the individuals and (b) sampling blog entries or social profiles written by those individuals. For the first part, the issues are similar to those facing survey and experimental researchers: How does one constitute a representative sample of the target population of interest? Generally, sampling techniques will vary depending on the variables of interest. For example, because the goal of Herring, Kouper, Scheidt, and Wright (2004) was to locate clear exemplars of the blog genre without focusing on any particular topic, they relied on a large blog tracking site at that time, blo.gs (now owned by Yahoo!), using the site's random selection feature.

But for other research questions, systematic random or stratified sampling may be the logical choice. For example, Mazur and Kozarian (in press) randomly selected blogs over a 4-month period by each research assistant alternating between six blog-hosting Web sites that news articles and preliminary searches had indicated as popular among the target age group. However, toward the end of the data collection period, stratified sampling was used to ensure appropriate representation for the age, gender, and racial groups that were included in the analyses. Stratification by day or month might be particularly appropriate for researchers of news content, who must tap daily and monthly variations in key variables (Neuendorf, 2002). In fact, cyclic variations are important to consider for all cases of sampling. For example, researchers analyzing blogs written by adolescents or by parents with school-age children would want to consider the influence of school vacations on topics discussed, and therefore relevant samples would include entries written at different times of the year.

When studying interactions between members of small blogging communities or social networks, snowball sampling might be appropriate. Efimova, Hendrick, and Anjewierden (2005) used this technique in their study of connections between blogs focusing on knowledge management and social software. It also may be appropriate when searching for blogs about topics infrequently discussed, such as uncommon illness or predicaments (e.g., New York City survivors of 9/11).

Once a particular blog is selected, researchers need to specify sampling guidelines to stipulate which entries (e.g., newest, oldest) and how many entries will be included for analysis, and whether selection of these entries will require originality, a minimum number of words or sentences, or a specific time interval. Some blog entries are often just a few lines and may include song lyrics, magazine quizzes, photographs, and other derivative or visually based material. Accordingly, Herring et al. (2004) excluded blogs with no text in the first entry, blogs that had not been updated within 2 weeks, and blog software used for other purposes to collect a sample composed solely of active, text-based blogs. Rodham et al. (2007), studying interpersonal interactions on a message board, collected every message posted during a full week and then all replies to those posts. Mazur and Kozarian (in press) purposefully skipped over first entries to allow time for readers to reply to adolescents' and emerging adults' blog narratives.

CONTENT CODING AND ANALYSIS

Once relevant social profiles or blog entries are located, I recommend that rather than code on the spot, you should print the documents and/or copy them into a Microsoft Word file, including user comments and links if desired. Human text coding seems to work better with hard copy, especially if it is helpful to write on it (Neuendorf, 2002). To preserve writers' anonymity, similar to practices of survey research, each social profile or blog should be assigned a number, but identifying information such as names, e-mail addresses, online handles, or links to personal Web pages should be omitted. However, because of its possible importance as a variable, dates of social profile update or blog or comment entry should be included.

In general, all measures for content analysis by human coders need to be delineated in a codebook, which corresponds with a coding form. A *coding form* provides spaces appropriate for recording the codes for all variables measure. Together, the codebook and coding form should work as protocol for content analyzing social profiles and blogs. See Neuendorf (2002) for an example of a paired codebook and coding form for measuring demographic and descriptive information. In fact, her book and that of Krippendorff (2004) are two comprehensive texts that are invaluable for this part of your analysis.

Clearly, coding categories will differ depending on research hypotheses, but social media analysis allows you to look specifically at some unique factors. This is because blogs and social network profiles are not only about content but are also about format. Consequently, although most researchers will want to begin their content analysis with *thematic coding*, which looks at the narrative as a whole for major issues, with social profiles and blogs, you can branch out beyond text to information in the profile itself, such as musical themes, visual images, and even the appearance of the Web page. You also can code for *affect*, which focuses on the feelings and needs expressed by the writer and which can be coded not only from emotion-related words and phrases but also from *emoticons* (i.e., graphical icons that represent emotions, such as the smiley face), music, and images. You can focus on *interaction coding*, which could include, at the simplest, consistency of posting. But the existence of public comments, links, trackbacks, and pingbacks allows you to quantify and code responses that flow between the initial social profiler and blogger and their readers and friends, as well as allows you to track other social profiles and blogs that may refer to them. Researchers of virtual worlds can code *avatars,* graphical icons that represent a real person in a virtual context, which can often be found on Web sites and game forums such as The Gamer Legion (http://forums.thegamerlegion.com); Second Life (http://www.secondlife.com), popular with adults; and Habbo (http://www.habbo.com), oriented to teens.

INTERCODER RELIABILITY

As in content analysis, coding is performed by measurement of content unit, in this case, as contained in a social profile or blog entry or by item. Traditionally, at least 10% of a random selection of all units is used for attainment of at least 80% intercoder reliability (Lacy & Riffe, 1996). Intercoder reliability is typically calculated as percent agreement: $R = 2 \ (C_{1,2})/ C_1 + C_2$. $C_{1,2}$ is the number of category assignments both coders agree on, and $C_1 + C_2$ is the total of category assignment made by both coders. This formula can be extended for n coders (Budd, Thorp, & Donohew, 1967). Other ways of calculating reliability, such as Scott's pi, Spearman's rho, or Pearson's r, can be found in Krippendorff (2004) and Neuendorf (2002), among other sources. Holsti's formula may be especially helpful if your data are nominal (Wimmer & Dominick, 2003).

Although the data before reconciliation are the data on which reliability is calculated, for data entry and analysis, coders must resolve discrepancies by relying on a consistently applied formal decision rule, such as the majority judgment prevails if there are three or more coders. Other ways to reconcile coding differences are by directing that the decisions of the primary coders are always retained or that coders must reach consensus in postcoding deliberations.

Summary and Conclusion

Social media are now part of mainstream social communication and information dissemination. As such, they are important to study in their own right. As research tools, they allow the researcher access to a large amount of text-based and visually based content written and responded to by international users more diverse than typical research samples. Social media analysis can be both enjoyable and infuriating, both for the principal investigator, astonished by much of the material telecast to millions by bloggers and social network users, and often for student research assistants, who are thrilled to flex their newly minted analytical skills on a medium with which they are so familiar but that, to a large extent, is presently almost unknown.

References

Ali-Hasan, N. F., & Adamic, L. A. (2007). *Expressing social relationships on the blog through links and comments.* Retrieved January 28, 2008, from http://www.noor.bz/pdf/ali-hasan_adamic.pdf

Ante, S. E. (2008, December 1). Facebook's land grab in the face of a downturn. *BusinessWeek,* p. 84.

Bahnisch, M. (2006). The political uses of blogs. In A. Bruns & J. Jacobs (Eds.), *Users of blogs* (pp. 139–149). New York: Peter Lang Publishing.

Budd, R. W., Thorp, R. K., & Donohew, L. (1967). *Content analysis of communications.* New York: Macmillan.

Calasanti, T. (2007). Bodacious berry, potency wood, and the aging monster: Gender and age relations in anti-aging ads. *Social Forces, 86,* 335–354.

Cathcart, R. (2008, January 10). MySpace said to draw subpoena in hoax case, newspaper reports. *Pittsburgh Post-Gazette,* p. A5.

comScore Media Metrix. (2005, August). *Behaviors of the blogosphere: Understanding the scale, composition and activities of weblog audiences.* Retrieved January 28, 2008, from http://www.comscore.com/Press_Events/Press_Releases/2005/08/US_Blog_Usage/%28language%29/eng-US

comScore Media Metrix. (2006, October 5). *More than half of MySpace visitors are now age 35 or older, as the site's demographic composition continues to shift.* Retrieved January 15, 2008, from http://www.comscore.com/Press_Events/Press_Releases/2006/10/More_than_Half_MySpace_Visitors_Age_35/%28language%29/eng-US

Dieu, B., & Stevens, V. (2007, June). *Pedagogical affordances of syndication, aggregation, and mash-up of content on the Web.* Retrieved January 15, 2008, from http://tesl-ej.org/ej41/int.pdf

Efimova, L. Hendrick, S., & Anjewierden, A. (2005, October). *Finding "the life between buildings": An approach for defining a weblog community.* Paper presented at Internet Research 6.0: Internet Generations, Chicago, IL. Retrieved January 22, 2008, from https://doc.telin.nl/dscgi/ds.py/Get/File-55092/AOIR_blog_communities.pdf

Goggin, G., & Noonan, T. (2006). Blogging disability: The interface between new cultural movements and Internet technology. In A. Bruns & J. Jacobs (Eds.), *Uses of blogs* (pp. 161–172). New York: Peter Lang Publishing.

Havenstein, H. (July 9, 2007). *Social networks becoming ingrained in daily adult life.* Retrieved January 15, 2008, from http://www.pcworld.com/article/134277/social_networks_becoming_ingrained_in_daily_adult_life.html

Herring, S. C., Kouper, I., Scheidt, L. A., & Wright, E. (2004). Women and children last: The discursive construction of Weblogs. In L. Gurak,

S. Antonijevic, L. Johnson, C. Ratliff, & J. Reyman (Eds.), *Into the blo-gosphere: Rhetoric, community, and culture of weblogs.* Retrieved January 22, 2008, from http://blog.lib.umn.edu/blogosphere/women_and_children.html

Hodkinson, P. (2006). Subcultural blogging? Online journals and group involvement among UK Goths. In A. Bruns & J. Jacobs (Eds.), *Uses of blogs* (pp. 187–197). New York: Peter Lang.

Huffaker, D. A., & Calvert, S. L. (2005). Gender identity and language use in teenage blogs. *Journal of Computer-Mediated Communication, 10.* Retrieved January 22, 2008, from http://jcmc.indiana.edu/vol10/issue2/huffaker.html

Krippendorff, K. (2004). *Content analysis: An introduction to its methodology* (2nd ed.). Thousand Oaks, CA: Sage.

Lacy, S., & Riffe, D. (1996). Sampling error and selecting intercoder reli-ability samples for nominal content categories. *Journalism & Mass Com-munication Quarterly, 73,* 963–973.

Lenhart, A., & Fox, S. (2006, July 19). *Bloggers: A portrait of the Inter-net's new storytellers.* Retrieved January 11, 2008, from http://www.pewinternet.org/PPF/r/186/report_display.asp

Lenhart, A., & Madden, M. (2007, January 3). *Social networking Web sites and teens: An overview.* Retrieved January 3, 2008, from http://www.pewinternet.org/~/media/Files/Reports/2007/PIP_SNS_Data_Memo_Jan_2007.pdf.pdf

Lenhart, A., Madden, M., Macgill, A. R., & Smith, A. (2007, December 19). *Teens and social media.* Retrieved January 3, 2008, from http://www.pewinternet.org/Reports/2007/Teens-and-Social-Media.aspx

Mazur, E., & Kozarian, L. (in press) Self-presentation and interaction in blogs of adolescents and young emerging adults. *Journal of Adolescent Research.*

McCoy, A. (2008, January 15). Social networking sites encourage Inter-net chat. *Pittsburgh Post-Gazette,* pp. C1–C2.

Mindlin, A. (2008, December 1). Maybe Canadians have more friends. *The New York Times,* p. B3.

Neuendorf, K. A. (2002). *The content analysis guidebook.* Thousand Oaks, CA: Sage.

Nizza, M. (2008, January 2). *Prankster playing Bhutto's son on Facebook fools news outlets.* Retrieved January 5, 2008, from http://thelede.blogs.nytimes.com/2008/01/02/prankster-playing-bhuttos-son-on-facebook-fools-news-outlets/

Oblinger, D. G., & Hawkins, G. L. (2006). The myth about putting infor-mation online. *Educause Review, 41,* 14–15.

Richards, L., & Mazur, E. (2008, March). *Adolescents' social networking online: Similar to or different from friendships offline?* Paper presented at the Society for Research in Adolescence Biennial Meeting, Chicago, IL.

Rodham, K., Gavin, J., & Miles, M. (2007). I hear, I listen, and I care: A qualitative investigation into the function of a self-harm message board. *Suicide & Life-Threatening Behavior, 37,* 422–430.

Schmidt, J. (2007). Blogging practices: An analytical framework *Journal of Computer-Mediated Communication, 12(4),* article 13, from http://jcmc.indiana.edu/issue4/Schmidt.html

Shrivastava, A. (2007, December 13). *Jangl launches SMS for Bebo's 40 million members.* Retrieved December 19, 2008, from http://internetcommunications.tmcnet.com/topics/broadband-mobile/articles/16466-jangl-launches-sms-bebos-40-million-members-worldwide.htm

Sifrey, D. (2007, April 5). *The state of the live Web, April 2007.* Retrieved January 17, 2008, from http://technorati.com/weblog/2007/04/328.html

Singer, J. (2005). The political j-blogger: "Normalizing" a new media form to fit old norms and practices. *Journalism, 6,* 173–198.

Skiba, D. J. (2007). Nursing education 2.0: Poke me. Where's your face in space? *Nursing Education Perspectives, 28* 214–216.

Stelter, B. (2008a, January 14). *MySpace agrees to youth protections.* Retrieved January 14, 2008, from http://www.nytimes.com/2008/01/14/technology/14cnd-myspace.html?_r=1&scp=1&sq=MySpace%20agrees%20to%20youth%20protections&st=cse

Stelter, B. (2008b, January 21). *From MySpace to YourSpace.* Retrieved January 21, 2008, from http://www.nytimes.com/2008/01/21/technology/21myspace.html

Stelter, B. (2008c, March 14). *AOL buying no. 3 social networking site.* Retrieved December 19, 2008, from http://www.nytimes.com/2008/03/14/technology/14aol.html?ref=technology

Stern, S. R. (2002). Virtually speaking: Girls' self-disclosure on the WWW. *Women's Studies in Communication, 25,* 223–253.

Stone, B. (2007, May 25). *Facebook expands into MySpace's territory.* Retrieved May 25, 2007, from http://www.nytimes.com/2007/05/25/technology/25social.html

Surridge, G. (2007, October 4). *37% of adults use social networking sites: Study.* Retrieved January 15, 2008, from www.financialpost.com/

Thiel, S. M. (2005). "IM me": Identity construction and gender negotiation in the world of adolescent girls and instant messaging. In S. R. Mazzarella (Ed.), *Girl wide Web* (pp. 179–201). New York: Peter Lang Publishing.

Williamson, D. A. (2007, December). *Social network marketing: Ad spending and usage.* Retrieved January 15, 2008, from http://www.emarketer.com/Reports/All/Emarketer_2000478

Wimmer, R. D., & Dominick, J. R. (2003). *Mass media research: An introduction.* Belmont, CA :Wadsworth.

Wollheim, P. (2007). The erratic front: YouTube and representations of mental illness. *Afterimage, 35*(2), 21–26.

Sonja Utz

Using Automated "Field Notes" to Observe the Behavior of Online Subjects

7

T his chapter deals with *online participant observation*. Online participant observation is the right method to study questions such as, How are conflicts solved in a Web forum or a chat? How do people form clans and coordinate cooperation in World of Warcraft? How and why do people trade objects in Second Life? After a short general introduction to the advantages and drawbacks of (online) participant observation, the chapter gives an overview of the various forms of participant observation. It explains what to observe in different online venues and, most important, how to automatize taking field notes in online participant observation. The chapter closes with a short overview of data analysis possibilities.

Participant observation is a useful method for researchers who want to study the online behavior of members of one or more online groups or virtual communities. Participant observation provides in-depth insights into the social relations and interpersonal processes within the observed group. It is an unobtrusive method, especially in the case of covert observation. Participants do not know that they are being observed and are therefore showing their natural behavior. This makes it the best method for studying sensitive issues (e.g., xenophobia and discrimination) or groups that will not deliberatively take part in research (e.g., criminals, members of hate groups, terrorists, viewers of porn sites). The method is also especially suited to the study of interpersonal interactions.

Participant observation is especially useful for generating hypotheses; the chapter is therefore useful for researchers who want to study new phenomena about which not much is known yet. *New phenomena* does not necessarily mean completely new phenomena genuine only to virtual settings, such as using magic in multiuser role-playing games or teleporting. Online participant observation can also address the question of whether well-known psychological principles apply also to virtual groups or how these principles are transformed in virtual settings. The chapter will be less relevant for researchers who want to use the Internet mainly as a tool for rapid data collection. Participant observation remains a time-consuming method, even if done via the Internet, and the results can often not be generalized to face-to-face interacting groups or even to other virtual communities.

Before starting the research, the researcher should also be aware of some other drawbacks to the method. Participant observation assesses only what can be observed, namely, behavior. Cognitions, motivations, and emotions get (partly) lost. In the case of covert observation, there are ethical issues to consider (see chap. 16, this volume). The generalizability of the findings to other groups or communities might be lower because the studied group is often a very specific one. In so-called going native lies the danger that the observer will identify too much with the participants and will no longer be able to interpret the data in an objective, scientific way. Participant observation is a time-consuming method—for observing as well as for coding and analyzing the data. Papargyris and Poulymenakou (2004), for example, spent 726 hours of active participation in a massively multiplayer online role-playing game (MMORPG), and Baym (1995) spent 2 years participating in the newsgroup rec.arts.tv.soaps.

Before turning to the actual implementation of the method, I first make some general remarks on (online) participant observation. Participant observation is a method mainly used by ethnologists, cultural anthropologists, and sociologists. The goal is the in-depth description of the studied group; the researcher neither focuses on a small set of variables and a large sample (as in online surveys; see chap. 12, this volume) nor imposes prefabricated categories on the subjects. Instead, the researcher tries to get a comprehensive picture of the life of the group. Ethnographers and sociologists use the method to study subcultures within society, for example, criminals, street gangs, or drug cultures. Participant observation is not only observation but also participation. Researchers spend a prolonged time in the observed group or community, months or even years (Schwandt, 1997), and immerse themselves in the (sub)culture.

With the rise of the Internet, new phenomena emerged, such as virtual communities. The first studies on virtual communities all used online participant observation as a method: Baym's (1995) study on newsgroups,

Curtis's (1992) study on multiuser dungeons (MUDs), Reid's (1991) study on Internet Relay Chat (IRC), and Rheingold's (1993) (less scientific) study of "The Well." The Internet is rapidly changing; new developments such as Second Life, social network sites (SNS), or YouTube can be studied with participant observation as well.

An important advantage of online participant observation is that researchers can stay in their office (Hine, 2000); hanging out with people who are addicted to drugs in dubious bars or on street corners or even moving to another country is no longer necessary. In groups with archived asynchronous communication, it is even possible to go back in time and download the old conversations, which would cut down research time as well. However, as Hine (2000) argued, "Part of following a newsgroup in real time is making sense out of the arrival of messages in the wrong order, waiting for responses to messages, and experiencing periods of high and low activity in the newsgroup" (p. 23). These experiences get lost if the researcher just downloads the archive.

Moreover, the Internet makes access to some groups easier. Subcultures such as those of sex workers or drug users are often especially difficult to access. It is often obvious that the researcher is not a member of the group, or the researcher has to change his or her appearance to gain the trust of the group members. Many factors that are a barrier for the study of certain groups, such as age, sex, and race/ethnicity are less apparent on the Internet. In general, online data collection is more unobtrusive than data collection in real-life participant observation. Field notes play a central role in traditional participant observation. If the researcher does not want to take notes conspicuously, he or she has to remember everything and write it down later. Participant observers often use audiotape recorders or digital cameras—but these devices are often visible, and people might not feel comfortable when being audiotaped or in front of a camera. Online participant observation can simply log conversations, take screenshots, or videotape the interaction on the monitor without anybody noticing it. Thus, the main advantages of online participant observation are its unobtrusiveness and its automatization of taking field notes.

Things to Think About Before Starting to Observe

Before starting the research project, the researcher has to think about which form of participant observation he or she wants to use. In general, this depends on the research question, the target group, and the technology used by the group. However, some factors are unique to online participant observation.

According to Lamnek (1995), forms of participant observation can be classified along several dimensions: the role of the observer (i.e., participatory vs. nonparticipatory), degree of participation (i.e., active vs. passive), transparency (i.e., overt vs. covert), directness of observation (i.e., direct vs. indirect), and standardization (i.e., structured vs. unstructured).

ROLE OF THE OBSERVER

Some observers participate fully in the lives of the people observed. Other observers remain outside, as pure observers. In most cases of traditional participant observation, the observer participates in the everyday lives of the people studied; the underlying assumption is that the processes and interaction can be better understood if the researcher gets directly involved (Jorgensen, 1989). Online participant observation is slightly different. First, it is often easier to remain in the complete observer role. In classical participant observation studies, the researcher who "just happens to" hang around all the time is noticed earlier than a lurker in a newsgroup. Second, participation in everyday life gets a different meaning because participation in the online groups is for most people only a part of their everyday lives. In many online groups, especially discussion forums or chats, there is relatively less "action" in which the observer can participate—for example, no playing jazz, no smoking marijuana (H. S. Becker, 1963), no taking off one's clothes at the nudist beach (Douglas, Rasmussen, & Flanagan, 1977). The main activity is communication, and it is often possible to understand this as a distanced observer.

In traditional participant observation, the role of the observer is often constrained. The observer is already part of the group, or he or she cannot become part of it (e.g., because he or she is of a different sex or race, or does not have the specific disease of the people being studied). Sometimes the researcher might not want to become a fully accepted group member, such as in the case of people who are racists or criminals. In general, it is easier to fully participate in online groups because some constraints, such as skin color, sex, and age, are not visible. However, group members might still recognize that the researcher is not really a member of the subgroup, for example, because the researcher does not understand and speak the jargon used in the group. It thus depends in part on the group studied whether participatory or nonparticipatory observation is the appropriate method. In case of nonparticipatory research, the researcher is a *lurker*—that is, he or she reads the discussions but does not contribute to them. As a participatory researcher, the researcher contributes to the group discussion or engages in other ways in the activities of the virtual group (e.g., plays the game in World of Warcraft, creates objects in Second Life). Whether to choose participatory or nonparticipatory observation depends also on the type of virtual group. Many discussions on Web forums can be read even if one is not a registered member of the community, and participation is therefore not inevitably necessary. However, in chats and multiuser environments such as Second Life or MMORPGs, the researcher has to at least create a profile or an avatar to get access to the community; nonparticipatory observation is not an option.

DEGREE OF PARTICIPATION

Closely related to the role of the observer is the dimension of active–passive participation. Gold (1958) distinguished between the complete observer (nonparticipatory observation), the observer as participant, the participant as observer, and the complete participant. In online communities, these stages could be related to the typology developed by Kim (2000), who distinguished members with regard to their involvement and amount of participation as visitors, novices, regulars, leaders, and elders. Thus, the participant observer could post occasionally to the discussion forum (i.e., a novice) or could become a regular poster (i.e., a regular). In MUDs, for example, the researcher could create a character and play a little bit but stay on a low level. However, the researcher could also advance many levels and become a so-called wizard—a member who is allowed to change the game by programming new areas and is also involved in decisions about the future of the community (i.e., a leader, in Kim's terms).

Active participation does not necessarily imply that the researcher changes the virtual world in disruptive ways. A researcher who writes controversial and provoking postings in a discussion forum probably disturbs the balance of the community; a researcher who plays World of Warcraft and advances several levels will probably not destroy the natural order.

The question of how far active participation biases objective observation has been discussed repeatedly. Jorgensen (1989) argued that objective and truthful findings are more likely if the researcher participates actively in the group because he or she gets firsthand access to what people think and feel. A too-distant researcher who sticks to his or her own cultural viewpoint might come to the wrong conclusions. The issue of the right degree of participation depends also on the research question. If the goal is to study the inner circle of community members or how wizards rule the virtual worlds, access to this subgroup (gained through active participation in the community) is often a necessary precondition to fully understand the occurring processes. If the goal is to study a discussion group, active participation is not inevitably necessary to understand the occurring processes. In general, the more complex the virtual world, the more necessary it is to learn not only the jargon but also how to live in this world (e.g., how to move, how to talk to each other, how to fight, how to trade), and more active participation is required.

TRANSPARENCY

Another question is whether the participant observation should be overt or covert (i.e., transparency). In the case of a passive observer, the observation will mostly be covert. However, an active participant observer has to choose whether to reveal that he or she is a researcher. Covert observation has an advantage in that it is more unobtrusive. The mere fact of knowing that one is observed can alter behavior (i.e., the Hawthorne effect), so covert observation has a better chance of detecting genuine behavior. However, covert observation raises ethical issues (see chap. 16, this volume), especially when it comes to the publication of the research results that describe the behavior or even quote statements of the people who were observed. Researchers using overt observation do not face this ethical problem. Overt observation also makes it easier to conduct the additional (informal) interviews that are often necessary to fully grasp a situation. In practice, it often turns out to be difficult to make sure that every community member knows that observation takes place. Daily posting might not be appreciated, and new members might not read all the old postings. In reality, it will often be the case that some members know that the researcher is an observer who does research and others do not. Nevertheless, the researcher should decide in the beginning whether to be open or to try to hide the researcher identity as long as possible.

DIRECTNESS

The dimension direct–indirect refers classically to how close to real life the observation is: Does the researcher observe the interacting people, or does the researcher afterwards analyze documents? This distinction becomes often blurry in online groups, for example, in case of discussion boards in which all actions are documented. However, as noted earlier, it makes a difference whether a discussion board or newsgroup is studied in real time, experiencing the pace with which messages arrive and feeling the dynamic, or whether all the threads are read and analyzed afterward. Thus, direct observation is recommended. However, direct observations can be complemented with the analysis of documents.

STANDARDIZATION

The dimension structured–unstructured refers to the degree of standardization and is mainly determined by the research question. In case of a specific research question and hypotheses, a structured approach is more suited, but for exploring a relatively new phenomenon, an unstructured approach is more suited. The choice of approach also depends on the stage of the project—most projects begin rather unstructured and become more and more structured (i.e., focusing on the sampling of specific events) in the course of the project.

In the Field

In the field, the most important rule is to try not to change the field. Especially for active overt observers, it is important to minimize possible researcher influences on the subjects. This might require adapting to the lifestyles of the observed subjects. This is often not so much a problem in online observation, but it is still important. The researcher has to be open to unexpected findings and ready to revise hypotheses. Depending on the subject, tolerance and acceptance of what is ordinarily unacceptable to most researchers might be necessary—for example, in studying so-called pro-ana Web sites, which promote an anorexic lifestyle as healthy and normal. It is also important to stay objective. However, the essence of participant observation is to try to see the situation from the perspective of the observed subjects, which requires a certain degree of perspective taking and identification with the subjects. However, in going native lies the danger of identifying too much with the subjects, which can lead to biased interpretation of the results. Thus, the researcher has to find a happy medium between these demands.

WHAT TO OBSERVE?

Observation begins from the moment the researcher enters the online group. The observation usually starts relatively unfocused; the goal is to become familiar with the setting. In traditional participant observation, this means focusing on the main features of the space or buildings. Jorgensen (1989) argued that the optimal situation would be to hang around without being noticed by anyone. This is relatively easy in virtual groups; even in worlds with avatars, it is quite common that an avatar is idle.

In the beginning, the researcher has to answer questions such as, Where are you? What people are there, and how many? and What are they doing? Mack, Woodsong, MacQueen, Guest, and Namey (2005) provided a list of aspects or categories researchers should observe, elaborated on what the respective category contains, and what the researcher should note. In Table 7.1, these categories are applied to the context of online participant observation.

In online participant observation, the setting can be simple in structure (e.g., discussion boards or the available chat channels). But even

TABLE 7.1

What to Observe During Online Participant Observation (Categories Based on Mack et al., 2005)

Category	Researchers should note
Appearance	Text-based chat: Nicknames or real names, kind of nickname Newsgroups: E-mail address with real name or anonymous e-mail, use of signatures Member profiles with pictures: Age, gender, clothing Avatars: Type of avatar (e.g., animal, comic, power, seductive)
Space and location	Web forums: Structure of the discussion boards (e.g., general, health, relationships), spatial metaphors (e.g., cafe, meeting room) 2D and 3D fantasy worlds: Space (e.g., building, forest, street corner), degree of realism (e.g., replication of existing space, not existing but realistic, fantasy world)
Verbal behavior and interactions	Newsgroups: Who initiates threads, who answers Chats, MUDs, and virtual worlds: Who speaks to whom how much; languages, dialect; use of emoticons, acronyms, and strategies to compensate the missing of nonverbal cues
"Physical" behavior and gestures	Virtual worlds (e.g., MUDs, 2D/3D chats, MMORPGs): Trading, fighting, dancing, and so on; what is part of the game (e.g., killing monsters in an MMORPGs); translation of "normal" behavior in virtual context or genuinely virtual behavior (teleporting); expression of emotions, social rank, social relationships
Personal space	2D and 3D worlds: How close people stand to one another in 3D worlds (measured in pixels apart via screenshots) or how close people associate with each other on SNS (top 8 on MySpace, networks of friends)
Human traffic	People who enter, leave, and spend time in the community (difficult for lurkers; otherwise often recorded in the logs)
People who stand out	People who receive a lot of attention: Popular members, older people, people with a high reputation, but also trolls (unwelcome intruders whose primary intent is to provoke and disrupt) and flamers (people who often harshly and rudely criticize others)

text-based MUDs are richly structured; they often consist of several continents with forests, lakes, meadows, deserts, mountains, caves, villages, and cities. Some of these worlds are replications of the real world or at least quite realistic worlds, but other MUDs are depicted on strange planets or other unfamiliar environments. The same holds true for graphical worlds—Is the graphical world a copy of some real-world place (e.g., Amsterdam in Second Life), or is it a genuine fantasy world? Avatars are important as well—What types of avatars do people use (Suler, 2007)? What do avatars do in this virtual world? Are they doing mundane things you could also do in physical life, such as talking, walking, drinking, eating, and dancing, or do they engage in behavior that would not be expected in physical life, such as killing a troll, healing other players, teleporting, creating objects, or practicing witchcraft? In what ways are these behaviors related to the goal of the virtual world or the game? In online participation in general, but especially in MUDs, MMORPGs, and virtual worlds such as Second Life, the distinction between in-character and out-of-character behavior becomes more important—How much does behavior reveal about the role of the avatar or the person behind it?

The problem of authenticity has been discussed in more detail by Hine (2000), that is, how can the researcher know how genuinely an avatar reflects its owner? The importance of authenticity depends also on the research question: Is the goal to find out how avatars behave in certain virtual worlds or to know how specific people behave in virtual worlds? What is observed is usually the behavior of an avatar or nickname, and the researcher has to be careful about drawing conclusions about the person behind them.

After becoming familiar with the general setting, the researcher's attention will generally turn to specific topics of interest. Those foci of attention will depend, of course, on the initial research question. If the researcher already has a specific question (e.g., What are the nature of greeting rituals?; B. Becker & Mark, 1998), the researcher can immediately focus on incidents related to the question. If the question is rather broad (e.g., How do communities emerge?; Papargyris & Poulymenakou, 2004), it is better to begin by observing a wide range of phenomena and then narrowing down the issue gradually. However, even with a specific question, it is important not to dismiss issues as irrelevant too quickly.

HOW TO TAKE FIELD NOTES

An observer must make detailed records, called *field notes,* of everything he or she hears and sees (Kidder & Judd, 1987): details on time, location, descriptions, other people, amount of noise, facial expression, and so on. Classical participant observers took notes during their observations, or in the case of covert observation, took notes no later than at the end of the day. Technology such as audiotape recorders or videotape recorders made it easier but also made the observation more obvious. Online observation does not entail these problems, yet the issue of how to take field notes in this environment remains. The simplest way might be to type the field notes immediately in a separate window while interacting in the virtual world. However, in online interactions this is often not necessary because the documents can be easily downloaded (mailing lists, newsgroups) or the interaction can be logged (e.g., chats, MUDs). How to save and archive the data depends on the type of communication (e.g., asynchronous vs. synchronous) and the underlying technology.

Chats and MUDs

For chats and MUDs, taking field notes corresponds to logging the sessions. IRC as well as Telnet clients usually have a log function. Sometimes it is a simple command such as "/log [on]." In the Windows-based client PuTTY, clicking on "Logging" under "Session" results in the screen displayed in Figure 7.1. Here, the researcher can easily specify whether to log only printable output or all session output, the name of the log file, and where it should be stored. The output is a .txt file that can be analyzed "by hand" or with text analyzing programs (see chap. 8, this volume). Chats using AOL Instant Messenger, MSN, or Google also have a log option. Programs have also been developed to allow the logging of chats in absentia. Originally intended for parents who want to monitor their children, these programs can be useful for research as well (for a link, see Additional Resources at the end of this chapter). Log files capture the text that appears on the screen of the person while he or she is logged in. In the case of indirect observation, logbots can be used. A LogBot is a sort of lurking participant—the bot stands in a room of the chat and logs all conversation, including time stamps, to a Web page or to a .txt file.

Newsgroups

Usenet discussions can be downloaded via the newsreader (included in most mail programs, or look at http://www.newsreaders.com/). There are online and offline newsreaders. Online newsreaders have to be connected to the Internet all the time, whereas offline newsreaders download all new articles to the hard drive. Thus, an offline newsreader should be chosen to archive posts.

Web Pages

Nowadays, many discussion boards are Web based, and often the members of the group can create profiles with pictures and additional information. The content of Web pages can be downloaded with the help of so-called offline readers, Web spiders, or Web crawlers. One example

FIGURE 7.1

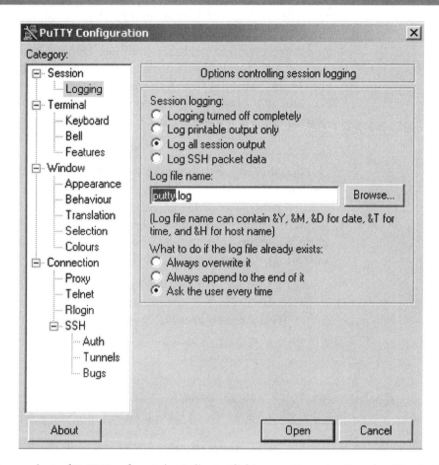

Screenshot of PuTTY, a free telnet client. Clicking on "Logging" under "Session" allows to specify options regarding the logging of the current session. From *PuTTY* [software package], by Simon Tatham, 2009. Copyright 1997–2009 by Simon Tatham. Reprinted with permission.

is HTTrack, a free Web spider. Downloading a Web site is easy and just takes a few steps (for a more detailed guide, see http://www.httrack.com/html/step.html):

1. Choose and enter a project name and a destination folder.
2. Choose an action and enter the URL. On the next screen, an action can be chosen. The default action is "download Web site," but it is also possible to download only specific file types or to update an existing download. Then the URL of the Web site has to be entered. Several filters, or download rules, can be specified

by clicking the "Options" button and choosing the respective tab. The researcher might want to exclude or include specific directories or domains, or download only sites containing a specific keyword. When mirroring large sites, for instance, SNS such as MySpace, it is important to set limits (see Figure 7.2) such as the mirroring depth. With a mirror depth of 4, the spider downloads the current pages and all pages it can get by clicking three times on any link. In MySpace, it would download all the pages of friends of friends of friends, although the researcher might be primarily interested in the discussion in a specific forum and the profiles of the contributors but not in their whole network. Another useful option is to enter login and password by using the "Add URL" button. This is important if the researcher wants to download a Web site that is only accessible to registered members.

3. It is possible to further specify the download (e.g., tell the program which provider to use, disconnect when finished, delay mirroring). Clicking on "Finish" starts the download.

In case of an error, error log files give additional information. Many possible problems are documented on the troubleshooting page, and the HTTrack Web page also hosts a forum for the discussion of problems.

FIGURE 7.2

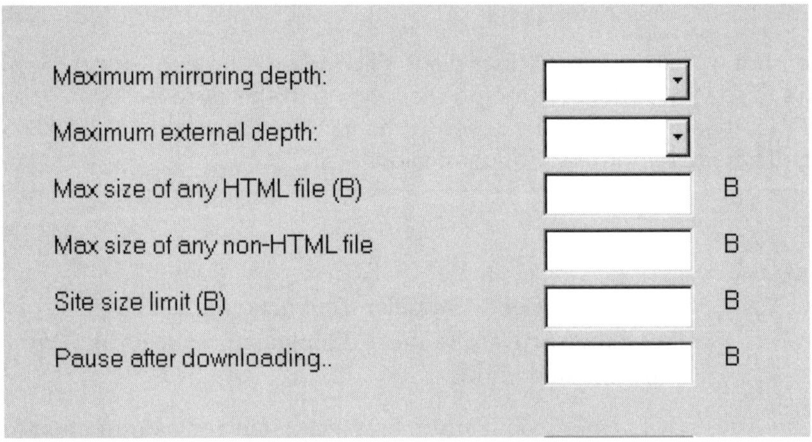

Maximum mirroring depth:

Maximum external depth:

Max size of any HTML file (B)   B

Max size of any non-HTML file   B

Site size limit (B)   B

Pause after downloading..   B

Screenshot of HTTrack. By clicking on "Set options" and choosing the tab "Limits," the researcher can specify the mirroring depths or other limits. From *HTTrack* [software package], by Xavier Roche & other contributors, 2009. Copyright 2009 by Xavier Rocher & other contributors. Reprinted with permission.

Social Networking Sites, Blogs, and Frequently Updated Web Pages

Chapter 6 (this volume) deals more specifically with the question of how to collect data from SNS and blogs. However, it should be briefly noted that many online venues such as SNS or blogs use feeds that can be easily used for data collection. SNS started to offer minifeeds to attend the users to changes in their network. Examples for such feeds are "*X* has changed her relationship status to single" or "*Y* has posted a new blog entry, 'Yes we can.' " At first, the feature's introduction caused some protests by Facebook users (Berger, 2006), but after privacy settings were added, the feature has become popular. This feature is now used by many other networks (e.g., Xing, LinkedIn, Hyves, Twitter) as well. Often these feeds can be received as RSS (Really Simply Syndication) feeds; that is, the researcher subscribes to the feeds and the RSS reader will check regularly for updates. RSS feeds do not afford much technological knowledge; they are integrated into most standard browsers, and all the researcher has to do is to click on the RSS symbol and on "subscribe to this feed." The RSS reader displays not only the updates but also metadata. Most blogs and many frequently updated Web pages have RSS feeds. Thus, RSS feeds are a fast and simple alternative to Web spiders when studying the communities around blogs.

However, RSS feeds also offer possibilities for studying SNS. Hyves, (www.hyves.nl) the largest Dutch SNS, offers feeds not only for the updates in their own network but also for the updates made by Dutch celebrities. The subscriber can choose the type of celebrity (e.g., actor, writer, politician) and the type of activity (e.g., relationships, reactions, providing new content). This technology was used, for example, in a study on the use of Hyves by politicians. Left-wing politicians were found to be more active on SNS then were right-wing politicians (Hyped, 2008). When examining SNS, feeds can be a useful supplement to Web spiders. Web spiders produce a huge amount of rather static data (e.g., profiles); feeds are better able to capture the dynamics of the interactions.

Graphical Worlds

Graphical worlds such as graphical chats, MMORPGs, or Second Life are more complex. However, the games often allow one to download the text messages (Peña & Hancock, 2006). Many of the avatar-based chats also have a log function. A frequently used system is the Active Worlds system. In the chat settings, the option to log the chat is provided. It is also possible to specify how many of the nearest avatars shall be fully rendered (see Figure 7.3)

Researchers can take screenshots and save them simply by pressing the "Print screen" button. A more sophisticated way would be to videotape the screen and analyze these videos later. However, this is

FIGURE 7.3

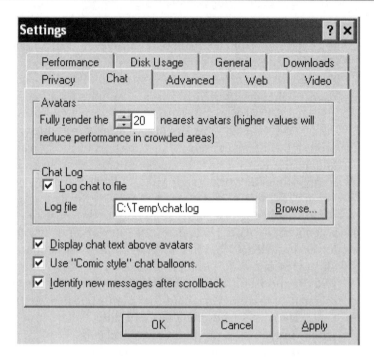

Screenshot of Active Worlds. Under "Settings" and "Chat," the number of nearest avatars that shall be fully rendered as well as the name and destination of the log file can be specified. From *Activeworlds* [software package], by Activeworlds, Inc., 2009. Copyright 1997–2009 by Activeworlds, Inc. Reprinted with permission.

usually not done. Remember, the goal of participant observation is not to fully capture every single action in the virtual world. The observation may not be too selective, but it is more important to understand the perspective of participants. Thus, it might be enough to take some screenshots of central avatars or of the rooms or areas on which the study focuses and to log exemplary chats or interviews.

Analyzing the Data

An important point is that—in contrast to quantitative research—analysis in qualitative research starts with the first observation. Data analysis and data collection are concurrent processes. The observation usually starts unstructured; then, recurring or otherwise interesting events are detected and initial hypotheses are developed. Additional observations are made, now more focused on the events directly relevant to the research question; the hypotheses are refined or revised, and so on. Consequently, the understanding the researchers have about the phenomenon influences further data collection (Thorne, 2000). Nevertheless, at the end of the project, researchers try to interpret the data as a whole. This analysis can be done quantitatively (see chap. 8, this volume), but most participant observers will choose a more qualitative approach.

Jorgensen (1989) described several different forms of theorizing that occur in participant observation: analytic induction, sensitizing concepts, grounded theory, existential theory, and hermeneutic (or interpretative) theory. Narrative analysis and discourse analysis focus on the analysis of language (Thorne, 2000) and can therefore be especially useful for analyzing chat and newsgroup discussions. The approaches differ in their exact procedure but usually start with coding and labeling the data. The concrete acts of behavior or communication described in the field notes have to be related to more abstract higher order categories. Essential features, recurring patterns, relationships between concepts, or the underlying structure have to be detected.

Many computer programs are available that support the analysis of qualitative data. Some types of data have to be prepared for data analysis. For examples, Hypertext Markup Language (HTML) tags have to be removed, and audiotaped or videotaped interviews have to be transcribed (see Additional Resources for examples). Programs for the actual analysis of data are atlas.ti, HyperRESEARCH V.2.06, MAXQDA, or NVivo7. Basically, these programs offer text search functions to search for specific topics; coding functions, which allows researchers to code (and recode, if necessary) the data; organizing functions to structure the project; writing tools that allow annotations and comments; and transfer of the output to other applications such as Word, Excel, or SPSS. It is beyond the scope of this chapter to explain these programs, but Lewins and Silver (2006) provided a useful guide that helps researchers to choose the right type of software for the type of project and data they have.

Conclusion

Online participant observation is a broad field because there are so many different online venues. It is not possible to cover all of the relevant aspects in one book chapter. The chapter gave an overview of this method and showed how issues of traditional participant observation translate into the online context. Moreover, it provided practical advice on using automated field notes. Although time consuming, online participant observation is still the best method to gain in-depth insight into social processes in virtual groups, and it is especially suited to explore relatively new phenomena.

Additional Resources

DATA COLLECTION

Logging MUDs
http://www.chiark.greenend.org.uk/~sgtatham/putty/
PuTTY–Telnet client with log option
Logging chats
LogBot: http://freshmeat.net/projects/logbot/
Chat Watch 5: http://www.zemericks.com/products/chatwatch/index.asp
Web spiders
Teleport Pro: http://www.tenmax.com/teleport/pro/home.htm (shareware, free trial version with reduced capacities)
HTTrack: http://www.httrack.com/ (free and well documented)
WebSpider 2: http://www.xaldon.de/products_webspider.html (free, German)
Web2Map: http://www.web2map.com/us/index.htm (shareware, first 30 days free, than $20)
GNU Wget: http://www.gnu.org/software/wget/wget.html (free utility for noninteractive download; e.g., works in the background while you are not logged in; no graphical interface, runs from command line; most features are fully configurable; for people with good programming skills)
MetaProducts Internet Collection (Offline Explorer Pro and Mass Downloader): Professional solution, reasonable price (about $100)

GENERAL TIPS FOR GETTING TEXT FROM WEB SITES

Linguistic Data Consortium tips for creating annotated data
http://www.ldc.upenn.edu/Creating/creating_annotated.shtml
Software for audio/video transcription
HyperTRANSCRIBE 1.5: http://www.researchware.com/ (free demo, license $99)

SOFTWARE FOR COMPUTER-AIDED QUALITATIVE DATA ANALYSIS

atlas.ti: http://www.atlasti.com/ (free trial license–includes all functions, without time limit, but limited to small projects; regular price $1,800; educational discount $585; student license $128)

HyperRESEARCH™ 2.8 http://www.researchware.com/(free demo version–no time limit, but limited to small projects; license $370; package with HyperTRANSCRIBE $399)

MAXQDA: http://www.maxqda.com/ (free trial version–30 day time limit, standard license $900, educational license $430, student license $125, leasing licenses possible)

NVivo: http://www.qsrinternational.com/products_nvivo.aspx (free trial–30 day time limit, full license for academics $595, student license $240)

Advene: http://liris.cnrs.fr/advene/index.html (free)

GENERAL OVERVIEW OVER QUALITATIVE RESEARCH

Qualitative research in information systems
http://www.qual.auckland.ac.nz/

References

Baym, N. K. (1995). The emergence of community in computer-mediated communication. In S. G. Jones (Ed.), *CyberSociety: Computer-mediated communication and community* (pp. 138–163). Thousand Oaks, CA: Sage.

Becker, B., & Mark, G. (1998). Social conventions in computer-mediated communication. In E. Churchill & D. Snowdon (Eds.), *Proceedings of CVE '98, Conference on Collaborative virtual environments* (pp. 47–56). England: University of Manchester.

Becker, H. S. (1963). *The outsiders.* New York: Free Press.

Berger D. (2006, September 9). Facebook gets the hint, adds privacy setting to News Feed. Retrieved August 8, 2008, from http://www.gadgetell.com/tech/comment/facebook-gets-the-hint-adds-privacy-settings-to-news-feed/

Curtis, P. (1992). *Mudding: Social phenomena in text-based virtual realities.* Retrieved January 15, 2007, from http://liquidnarrative.csc.ncsu.edu/classes/csc582/papers/curtis.pdf

Douglas, J. D., Rasmussen, P. K., & Flanagan, C. A. (1977). *The nude beach.* Beverly Hills, CA: Sage.

Gold, R. L. (1958). Roles in sociological field observations. *Social Forces, 36,* 217–223.

Hine, C. (2000). *Virtual ethnography.* London: Sage.

Hyped. (2008). *Vooral linkse politici actief op Hyves* [Mainly left-wing politicians active on Hyves]. Retrieved November 6, 2008, from http://www.hyped.nl/details/20080502_vooral_linkse_politici_actief_op_hyves/

Jorgensen, D. L. (1989). *Participant observation: A methodology for human studies*. Newbury Park, CA: Sage.

Kidder, L. H., & Judd, C. M. (1987). *Research methods in social relations* (5th ed.). New York: CBS Publishing.

Kim, A. J. (2000). *Community building on the Web: Secret strategies for successful online communities*. Berkeley, CA: Peachpit Press.

Lamnek, S. (1995). *Qualitative Sozialforschung: Band 2. Methoden und Techniken*. [Qualitative research in social science: Vol. 2. Methods and techniques]. Weinheim, Germany: Psychologie Verlang Union.

Lewins, A., & Silver, C. (2009). *Choosing a CAQDAS package* [Working paper]. Retrieved March 8, 2007, from http://caqdas.soc.surrey.ac.uk/PDF/2009%20Choosing%20a%20CAQDAS%20Package.pdf

Mack, N., Woodsong, C., MacQueen, K. M., Guest, G., & Namey, E. (2005). *Qualitative research methods: A data collector's field guide. Family Health International*. Retrieved March 12, 2007, from http://www.fhi.org/NR/rdonlyres/ed2ruznpftevg34lxuftzjiho65asz7betpqigbbyorggs6tetjic367v44baysyomnbdjkdtbsium/participantobservation1.pdf

Papargyris, A., & Poulymenakou, A. (2004). Learning to fly in persistent digital worlds: The case of massively multiplayer online role-playing games. *SIGGROUP Bulletin, 25*, 41–49.

Peña, J., & Hancock, J. T. (2006). An analysis of socioemotional and task communication in online multiplayer video games. *Communication Research, 33*, 92–102.

Reid, E. (1991). *Electropolis—Communication and community on Internet Relay Chat*. Retrieved January 7, 2000, from http://www.aluluei.com/electropolis.htm

Rheingold, H. (1993). *The virtual community: Homesteading on the electronic frontier*. Reading, MA: Addison-Wesley.

Schwandt, T. (1997). *Qualitative inquiry: A dictionary of terms*. Thousand Oaks, CA: Sage.

Suler, J. (2007). *The psychology of avatars and graphical space in multimedia chat communities*. Retrieved March 9, 2007, from http://www.rider.edu/~suler/psycyber/psyav.html

Thorne, S. (2000). Data analysis in qualitative research. *Evidence-Based Nursing, 3*, 68–70.

Matthias R. Mehl and Alastair J. Gill

Automatic Text Analysis

8

This chapter provides an introduction to the application of automatic text analysis (ATA) in online behavioral research. We use the term *ATA* synonymously with the terms *computer content analysis, computer-assisted content analysis, computer-assisted text analysis,* and *computerized text analysis. ATA* has been defined as a set of methods that automatically extract statistically manipulable information about the presence, intensity, or frequency of thematic or stylistic characteristics of textual material (Shapiro & Markoff, 1997). In line with a quantitative notion of measurement, we focus exclusively in this chapter on ATA tools that extract quantitative information that can be subjected to statistical analysis.

The chapter covers basic information that helps researchers identify how they can use ATA in their online research. To maximize the chapter's utility, we focus on two specific ATA tools: Linguistic Inquiry and Word Count (LIWC; Pennebaker, Francis, & Booth, 2001) and Wmatrix (Rayson, 2008). We selected these tools because they (a) cover a range of ATA needs, (b) are user friendly and operate fully automatically, and (c) are maintained by researcher groups with a track record in the field. We also selected them because we have used them extensively in our own online research (Cohn, Mehl, & Pennebaker, 2004; Gill, French, Gergle, & Oberlander, 2008; Lyons, Mehl, & Pennebaker, 2006; Oberlander & Gill, 2006).

Broader reviews of ATA strategies are provided by Mehl (2006), Krippendorff (2004), Neuendorf (2002), Popping (2000), and West (2001).

What Is the Value of Automatic Text Analysis for Online Behavioral Research?

ATA is a valuable method for online research for at least three reasons. First, textual data, the input for ATA, is abundant on the Internet. World knowledge is increasingly available online. Collaborative enterprises such as the Google Library Project or the Open Content Alliance are creating full-text online indices of millions of digitized documents. Furthermore, within only a few years, the Internet has become an indispensable means of daily communication. People routinely interact with others through e-mail, instant messages, chat rooms, blogs, and social networking sites. From a researcher's perspective, such text-based Internet data are informative and can be used to study psychosocial phenomena without running actual participants (for ethical considerations around the use of Internet data in research, see chap. 16, this volume).

For example, national newspaper coverage can be analyzed to compare the prevalence of psychological themes across cultures. Similarly, people's responses to disasters can be studied through the tracking of public postings on social sharing Web sites (Stone & Pennebaker, 2002). The global and archival nature of the Internet has made it possible to simulate the virtual equivalent of a multisite, longitudinal study conveniently and retroactively from the investigator's office computer—with the opportunity to obtain extensive baseline information on unpredictable events such as disasters after the fact (Cohn et al., 2004).

Second, textual data are often collected as part of online research anyway. ATA, then, can provide additional, low-cost means for exploratory data analysis. Researchers routinely include open-ended questions in their online surveys. Because of the burden of manual coding, however, participants' answers to these questions often remain unanalyzed. ATA can efficiently content analyze free responses. For example, in a Web-based survey of responses to the attacks of September 11, 2001, many participants responded to the final open-ended question "Is there anything else you would like to add?" in considerable detail and provided their accounts of the events (Skitka, L. J. personal communication, August 15, 2006). An ATA of cognitive complexity in such stories could help reveal individual differences in the processing of traumatic life events.

Third, data derived from ATA have some unique psychometrically desirable features: (a) They share zero method variance with the most common method in the social sciences, the self-report rating scale; (b) they are objective in the sense that they ensure measurement equivalence across studies and labs using the same tool; (c) they are expressed in a nonarbitrary, naturally meaningful metric, the number or percentage of words in a text that fall into a certain category (e.g., positive emotion words, adverbs). These psychometric features positively affect the generalizability and ecological validity of text-analytically derived findings.

What Are the Potential Limitations of Automatic Text Analysis for Online Behavioral Research?

ATA has the following potential limitations. First, it can be somewhat inflexible in its application. Whereas questionnaires can be constructed to measure any construct, ATA is generally constrained by the variables that the programs provide. LIWC (Pennebaker et al., 2001), for example, has standard categories for positive and negative emotion words but does not extract information about specific emotions, such as pride, shame, or guilt. Similarly, Wmatrix (Rayson, 2008) identifies different types of verbs but not the ones suggested by the Linguistic Category Model (Semin & Fiedler, 1988). However, some programs do provide users with limited freedom over the analysis. LIWC, for instance, can search for lists of target words through user-defined dictionaries. Cohn et al. (2004) used this option to count how often participants used words such as *Osama, terrorist,* or *hijack* in their blogs after September 11, 2001.

Second, ATA applications are not always designed with the needs of the average behavioral scientist in mind. Some of the more powerful tools have been developed within computational linguistics and artificial intelligence and have their primary application in these fields (e.g., Coh-Metrix [Graesser, McNamara, Louwerse & Cai, 2004]; Latent Semantic Analysis [Landauer, Foltz, & Laham, 1998]). Yet these tools can be successfully used to answer behavioral research questions, and because of their computational advantages, they often extract critical language information that is lost with a simple word count (e.g., Campbell & Pennebaker, 2003). The consequence of using these tools outside of their original domain tends to be a loss of user friendliness. For example, Wmatrix, an application developed in corpus linguistics, does not have the "each-participant-a-line–each-variable-a-column" setup that psychologists are used to. Instead, it operates by comparing two text corpora (Oberlander & Gill, 2006; Rayson, 2008).

Finally, (word-count–based) ATA is sometimes viewed as simplistic in its approach. Mehl (2006) provided an in-depth discussion of (and rebuttal to) this concern. In essence, most word-count–based ATA tools (e.g., LIWC) neglect grammar (e.g., they do not distinguish between "the mother yelled at her child" and "the child yelled at her mother"); confuse context-specific word meanings (e.g., "What you did made me *mad*" vs. "I am *mad* about your cute curls"); and take metaphors (e.g., "I am on cloud nine"), irony (e.g., "It was as pleasant as getting a root canal"), and sarcasm (e.g., "Thanks a lot for blaming me for this") for their literal meanings. More sophisticated tools, such as Wmatrix, are beginning to address these issues. However, it is important to note that it is not the computational sophistication of an ATA tool that determines the validity of a text-analytically derived finding; it is the degree to which the extracted linguistic information unambiguously answers a research question. For example, for the question of whether self-focused attention in depression manifests itself in an elevated use of first-person singular, the specific context in which people with depression use *I*, *me*, and *my* is not immediately relevant (Chung & Pennebaker, 2007).

In the remainder of the chapter, we provide a user guide for analyzing text-based Internet data with ATA. Because of space constraints, we limit this user guide to two ATA tools: LIWC (Pennebaker at al., 2001), as a word-count–based program that has gained considerable popularity within psychology; and Wmatrix (Rayson, 2008), as a more complex, Web-based ATA application developed in corpus linguistics. We illustrate the steps involved in a LIWC and Wmatrix analysis using (slightly modified) excerpts from four daily blogs. The blogs were selected from a larger data set collected by Nowson (2006). Two of the four blogs were written by female students (Blogs A and B) and two, by male students (Blogs C and D). The excerpts of the sample blogs are shown in Exhibit 8.1. In our user guide, we aim at providing sufficient detail to allow researchers to start analyzing their own textual data after reading the chapter. We also supplement our step-by-step guide with recommendations based on our own experiences working with the two tools.

Word-Count–Based Psychological Text Analysis: Linguistic Inquiry and Word Count

LIWC was developed in the 1990s in the context of research on the salutary effects of writing about traumatic experiences. Over time, LIWC has been used more broadly to study the psychological implica-

EXHIBIT 8.1

Excerpts From Four Sample Blogs

Blog A (Female Author 1)
Imagine you are happy and life is good, but it wasn't always like that, you once had a love and, at that point you knew, you felt that this was the one. But he wasn't. He broke you, he took everything you had and more from you, but you gave to him because you couldn't have enough of him. Then one day you realized how bad he was and you broke away. It took you a long time to get away from him, he had tied up your emotions and controlled you physically, you knew that he didn't want it to end.

Blog B (Female Author 2)
I find that I just don't have the stamina or brain power to write about war or politics at night. Which is fine, because you can't be all vitriolic and indignant 24 hours a day. Well you can, but I have better things to think about. I was thinking about writing more. No, not blogging more. Writing more. I was thinking about finding an agent. I was thinking of how I wanted to be nothing but a writer since I was about seven, how holding a pencil in my hand at that young age made me feel alive and important.

Blog C (Male Author 1)
This day fucking blows. I mean, I'm really happy for all the seniors, they're finally free, but this sucks for me. Everyone who I could really talk to is gone. Bye C**, bye N**, bye even M**, bye everyone else. I'll probably stay in touch with most of them, so that's not even the worst part. If you know me you probably know what other horrible connotation this day has for me. I always knew it was coming, but could never really believe it. Right now I'm completely crushed. Then all these other stupid things popped up to make the day worse.

Blog D (Male Author 2)
I can't take it anymore, I'm going absolutely crazy. I'm locked up in this house and haven't been out besides school in a week. I'm questioning myself. And I'm totally fucking obsessed. Even had a falling out with someone today, and I didn't want that to happen at all. And I just don't know what to do. There's nothing I can do, it's just building inside me and has no way to escape. I feel like I'm going to explode. The only time I feel good is right after a run, and I fucking suck and can't even run a lot anymore.

Note. The excerpts were spell checked; the original punctuation was maintained; asterisks were used to preserve the authors' confidentiality. Data from *The Language of Weblogs: A Study of Genre and Individual Differences,* by S. Nowson, 2006, Unpublished Doctoral Dissertation, Scotland: University of Edinburgh. We are grateful to Scott Nowson for kindly making these data available.

tions of language use (Pennebaker, Mehl, & Niederhoffer, 2003). LIWC is a word-count–based ATA tool that operates by comparing each word of a given text with an internal dictionary consisting of 2,300 words. The default LIWC dictionary comprises 74 grammatical and psychological dimensions. The LIWC program was revised in 2001 and recently underwent a second major conceptual and technical revision (Pennebaker, Booth, & Francis, 2007). Because of the high popularity of, and our extensive experience with, LIWC 2001, we provide our user guide for this version. More information on LIWC 2007 is available at http://www.liwc.net.

LIWC is one of the most widely used ATA tools in psychology (Mehl, 2006). Its popularity is in part due to its effectiveness in meeting the needs of behavioral scientists: (a) it analyzes basic grammatical features of texts but also provides information about important psychological processes; (b) its categories have been psychometrically tested (c); the software is extremely user friendly; (d) it is available in several languages, and translated dictionaries with demonstrated equivalence to the original English dictionary are available in German (Wolf et al., 2008), Spanish (Ramírez-Esparaza, Pennebaker, García, & Suriá, 2007), and Dutch (Zijlstra, van Meerveld, van Middendorp, Pennebaker, & Geenen, 2004) (psychometrically untested translations exist for Italian, Norwegian, and Portuguese, and new translations are being developed for Chinese, Hungarian, Korean, Polish, Russian, and Turkish); and (e) numerous studies have successfully used LIWC and thereby contributed to the construct validity of its categories.

PREPARING THE DATA FOR LINGUISTIC INQUIRY AND WORD COUNT ANALYSIS

Data collection generally starts with sampling online text from participants (e.g., e-mails, instant messages) or directly from the Internet (e.g., blogs, chat rooms, discussion boards). We recommend the following steps to prepare the data for LIWC analysis:

1. Save the collected texts as plain text files (.txt); even though LIWC can analyze delimited text segments, we recommend creating a separate text file for each unit of analysis (i.e., each personal home page or each daily blog in a nested design).

2. Clean the texts by applying the "what you see is what LIWC gets" rule; remove any word that does not reflect the author's language (e.g., e-mail histories, signatures, system information, advertising, buttons).

3. To maximize word recognition by the dictionary, submit the texts to an automatic spell-checker (LIWC analyses are case insensitive). Note, however, that Wolf et al. (2008) recently demonstrated that LIWC analyses of longer texts (more than 400 words) are fairly robust against regular amounts of typos and misspellings.

4. Render the texts consistent with LIWC typing conventions (documented in the program's "Help" menu). These conventions regulate the handling of colloquialisms (e.g., *gotta*) and abbrevi-

ations (e.g., *w/*). Common verb contractions are included in the dictionary and need not be changed (e.g., *I'm, we're, don't, isn't*). The use of slang in e-mails, chats, and instant messages (e.g., *LOL, CUL*) can challenge LIWC. If use of slang is of interest, user-defined dictionaries should be created to capture the relevant words or abbreviations. Otherwise, slang or abbreviations should be spelled out.

5. Manually tag fillers in natural language (see also the program's "Help" menu). To avoid misclassifications, change *well, like, you know, I mean,* and *I don't know* to *rrwell, rrlike, youknow, Imean,* and *Idontknow* where they are used to fill speech gaps (e.g., *I mean, I didn't like rush* to *Imean, I didn't rrlike rush*). The tagging is facilitated by using a "Search and Replace" function. However, it is critical to search for each individual occurrence and not to use "Replace All."

6. For confidentiality reasons, remove personally identifying information; using asterisks (e.g., ***) helps to keep the word count accurate.

RUNNING THE LINGUISTIC INQUIRY AND WORD COUNT ANALYSIS

Running the data through LIWC is straightforward. Clicking "Process text" in the "File" menu opens the "Select file(s) to process" window. Clicking "Select" after the text files to be analyzed have been marked opens the "LIWC results file" window where the output file is specified; clicking "Save" runs the analysis; and the output file ("LIWC results.dat") opens after all files are processed. As a hands-on example, we submitted the four sample blogs depicted in Exhibit 8.1 to a LIWC analysis (see Table 8.1). Each blog was saved as a separate text file, cleaned, and spell checked. *I mean* in Blog C was tagged as a filler (*Imean*) and names were deidentified (e.g., *C***). No other manual changes were made. Figure 8.1 shows a screenshot of the LIWC analysis.

We recommend processing text files from a folder that is located at the level below the hard drive (e.g., D:\LIWCtemp). Because of a bug in some versions of the software, LIWC 2001 sometimes crashes (i.e., freezes) when it processes files that are stored lower in the data hierarchy.

INTERPRETING THE LINGUISTIC INQUIRY AND WORD COUNT OUTPUT

The LIWC output is a tab-delimited text file that can be imported into statistical software packages. Each column contains one LIWC variable; each row, the language information for one text file. The first column

TABLE 8.1

Results of the Linguistic Inquiry and Word Count (LIWC) Analysis of the Four Sample Blogs

LIWC variable	Blog A	Blog B	Blog C	Blog D
Raw word count	102.0	100.0	101.0	102.0
Words captured by the dictionary	89.2	80.0	74.3	89.2
Emotional processes				
Emotion words	4.9	1.0	7.9	5.9
Positive	3.9	1.0	2.0	1.0
Negative	1.0	0.0	5.9	4.9
Cognitive processes				
Words of more than six letters	7.8	15.0	12.9	15.7
Cognitive mechanism words	12.8	14.0	9.9	3.9
Causation words	2.0	4.0	0.0	0.0
Insight words	3.9	6.0	4.0	2.0
Interpersonal processes				
First person singular pronouns	0.0	10.0	7.9	13.7
First person plural pronouns	0.0	0.0	0.0	0.0
Second person pronouns	14.7	2.0	2.0	0.0
Third person pronouns	8.8	0.0	2.0	0.0
Social words	25.5	2.0	7.9	2.0

Note. All LIWC variables except raw word count are expressed in percentages of total words; for Blog C, *I mean* was manually changed to *Imean* to tag it as a filler word; for heuristic purposes, the selected LIWC variables have been arranged into three important psychological domains.

shows the file name; the first row, the LIWC variable names. All variables (except the raw word count) are expressed in percentages of total words and are thus controlled for text length. LIWC by default provides language information along 74 dimensions.

The four blogs were comparable in length (see Table 8.1); in consideration of the reliability of low base-rate categories, we generally recommend using texts of at least 100 words. Across the four blogs, LIWC recognized around 80% of the words, which is typical for nontechnical language. Because statistical analyses with an *N* of 4 are not meaningful, we compare the blogs descriptively with regard to selected variables that have repeatedly been found to be implicated in psychological processes (Chung & Pennebaker, 2007). Heuristically, these variables can be thought of as capturing three important psychological domains: emotional processes (positive and negative emotion words), cognitive processes (cognitive mechanism words, words of more than six letters), and interpersonal processes (personal pronouns, social words). This is not to suggest that the LIWC categories that fall outside of these three domains are not important. Naturally, it is the research question that determines the relevance of a specific LIWC variable (e.g., sexual words,

FIGURE 8.1

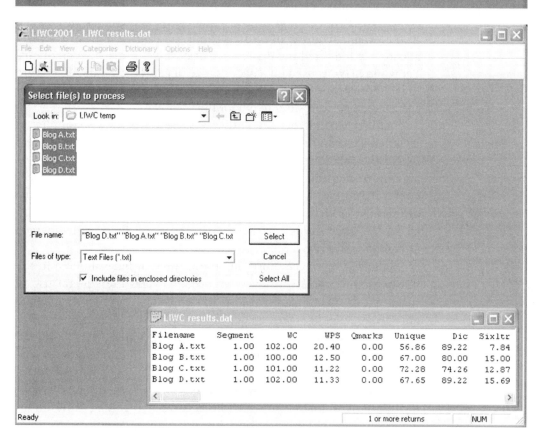

Screenshot of the Linguistic Inquiry Word Count (LIWC) analysis of the four sample blogs; clicking "Select" in the "Select file(s) to process" window opens the "LIWC results file" window, where the name of the output file is specified; clicking "Save" runs the analysis, and the tab-delimited output file ("LIWC results.dat") opens after all files have been processed. From *LIWC* [Computer Software], by W. Pennebaker, Roger J. Booth, and Martha E. Francis, Pennebaker Conglomerates. Copyright 2007, LIWC.net, Pennebaker Conglomerates. Reprinted with permission.

numbers); within our own research, however, these categories have repeatedly emerged as important (Mehl, 2006).

With regard to emotional processes, Blog A emerged as relatively positive in emotional tone (3.9% positive vs. 1.0% negative emotion words), Blog B as neutral (1.0% vs. 0%), and Blogs C and D as quite negative (2.0% vs. 5.9%, and 1.0% vs. 4.9%, respectively); this is consistent with our impression after reading the blogs. Note that, in our opinion, the labels of the LIWC emotion categories can be slightly misleading. "Positive emotion words" includes words such as *careful* and *perfect;* "negative emotion words" includes words such as *doubt* and *fail.* In the original context of writing about a trauma, these words likely adequately indicated the experience of positive or negative emotions. Yet when LIWC is used on a wider range of genres, the categories seem to tap more generally into the emotional tone of a text rather than the specific verbal expression of emotions.

With regard to cognitive processes, Blog A was less complex in its language than the other three blogs (only 7.8% of the words were longer than six letters). Blogs A and B, however, contained considerably more cognitive mechanism words (12.8% and 14.0%, respectively) than Blogs C (9.9%) and D (3.9%), suggesting a higher degree of cognitive (self-)reflection. It is interesting to note that Blog B, with its existential concerns and slightly esoteric word choice, also emerged as highly cognitive overall.

Finally, and maybe most important, the four blogs differed in the interpersonal processes they referenced. Beyond the general use of social words such as *talk* or *share,* interpersonal processes tend to be encoded in language through personal pronouns. Whereas Blog A was written from a detached second person perspective (*you;* 14.7%), Blogs B and D used first person singular at a high rate (10.0% and 13.7%, respectively). The frequent use of *I, me,* or *my* indicates personal involvement, with attention being on the self as social reference point. Psychologically, use of first person singular is correlated among other things with vulnerability for depression, low self-esteem, and the experience of stress (Pennebaker et al., 2003). Consistent with the psychological urgency it conveys, Blog D came out highest in the use of first person singular.

In sum, our analyses show that LIWC extracts language information at a psychologically meaningful level. This information often converges with ad hoc impressions derived from reading a text but also goes beyond what is noticed by a human observer (Mehl, 2006).

ADVANCED LINGUISTIC INQUIRY AND WORD COUNT FEATURES

LIWC has a few advanced features, such as the option of loading other dictionaries (e.g., foreign language dictionaries, special pronoun or particle dictionaries), creating user-defined dictionaries, and analyzing text in segments. Because of space limitations, we refer the interested reader to the detailed information provided in the program's "Help" menu and the LIWC manual.

Advanced Grammatical and Semantic Category Analysis: Wmatrix

Wmatrix is a powerful, Web-based ATA tool that uses corpus linguistics methods (Rayson, 2008; note that our description is based on Wmatrix2, the most recent version of the program). Corpus linguistics studies language using a large (usually electronic) collection of texts (i.e., a *corpus*; McEnery & Wilson, 1996). The research question often determines which texts are included in the corpus, whether they represent the whole of the English language (e.g., the British National Corpus; BNC; Burnard, 1995), or a particular group of interest (e.g., English language learners, International Corpus of Learner English; ICLE; Granger, Dagneaux, & Meunier, 2002). In addition to working with existing corpora, researchers also use texts that were collected under specific conditions (e.g., for a study on personality and language use; Oberlander & Gill, 2006).

Analytically, Wmatrix uses a *corpus comparison* approach, which compares a corpus of interest, the *Research Corpus*, to a second corpus, the *Reference Corpus*. This process identifies ways in which the research corpus differs from the reference corpus. For example, to examine characteristics of second language learners, a research corpus such as the ICLE may be compared against a general collection of English language, for example, the BNC. Alternatively, corpora derived from two comparable groups (e.g., male and female authors) may be compared.

Wmatrix and LIWC differ in two main ways. First, Wmatrix does not impose a set of relevant language features (defined by the dictionary). Instead, it extracts comprehensive word use, grammatical, and semantic information on a set of texts. Data-driven approaches like Wmatrix operate bottom-up because they allow characteristic language features to emerge from the data. Dictionary-based ATA tools like LIWC, in contrast, operate top-down by focusing on predefined, theoretically derived dictionaries. Second, Wmatrix is more sophisticated than most dictionary-based ATA tools. Wmatrix uses an automatic part-of-speech tagger (Constituent-Likelihood Automatic Word-Tagging System [CLAWS] tagger) to disambiguate and classify the syntactic function of words in a sentence (e.g., in "She is a mine worker," *mine* is a noun, not a pronoun). Similarly, Wmatrix uses a semantic tagger (UCREL Semantic Analysis System [USAS] tagger) to automatically disambiguate and classify the semantic function of words in a sentence (e.g., Wmatrix codes the example of *cloud nine* as "happy," rather than "meteorological").

Yet Wmatrix can also create challenges for behavioral researchers. First, the detailed information at the individual word level can be difficult to interpret. What does it mean if women overuse the words *about* and *was* relative to men? This problem can be reduced by focusing on Wmatrix's analysis of broader grammatical and semantic features. With this type of analysis, Wmatrix results (e.g., use of "superlatives") can be interpreted similar to LIWC results (e.g., use of "emotion words") with the difference that they are based on more comprehensive linguistic information.

Second, as noted before, Wmatrix analyzes text at the corpus level, not at the level of the individual author. Therefore, data collected from participants need to be clustered into text corpora. Although clustering texts is straightforward with discrete variables (e.g., gender, disease diagnosis, experimental group), it is more complex with continuous variables (e.g., age, extraversion). Then, the data need to be categorized, for example, by splitting a variable on the median or by forming extreme groups (e.g., participants low vs. high in extraversion; Oberlander & Gill, 2006). In this way, Wmatrix has been used to examine attitudes toward fashion (Wilson & Moudraia, 2006) and to code judgments of language used by science learners (Forsyth, Ainsworth, Clarke, Brundell, & O'Malley, 2006).

PREPARING THE DATA FOR WMATRIX ANALYSIS

We now demonstrate a Wmatrix analysis step-by-step using our four sample blogs. Steps 1 through 4 of the data preparation are identical to those described in the section "Preparing the Data for Linguistic Inquiry and Word Count Analysis" (save as plain text files; clean up extraneous text not written by the author; check spelling; carefully consider the inclusion of abbreviations or slang). The following steps, however, are unique to preparing texts for analysis with Wmatrix:

5. Wmatrix recognizes normal alphanumeric characters (*A–Z, a–z, 0–9*), but special characters (e.g., %, &, *, ', ") require care and should be encoded using Standard Generalized Markup Language (SGML; e.g., ampersand, &, is encoded as "&amp"). The use of punctuation also needs consideration; for details, see http://www.comp.lancs.ac.uk/ucrel/claws/format.html.

6. For Wmatrix, it is recommended to render data anonymous by substituting alternative names for the original ones because, for example, asterisks as a means of deidentification ('**') are not recognized by the program.

7. Because text corpora are built from the original texts by merging the individual text files into larger files, we recommend including a unique identifier at the start and end of each text within a corpus; for example the start of a single text can be marked as "<text=filename>" and the end as "<\text=filename>" (with "filename" as the identifier).

In addition to plain text (.txt) files, Wmatrix also supports HTML format as input files, which is particularly useful when dealing with online content (and which we recommend when the data contain SGML characters). A useful feature of Wmatrix is that it lists words it could not classify; we recommend checking this list for spelling or typographical errors. Note that fillers (e.g., *well, like*) need not be manually tagged because Wmatrix detects them automatically.

Finally, corpora need to be built from the individual texts. In our example, we cluster the four blogs into blogs by female authors (Blogs A and B) and male authors (Blogs C and D) and merge the four text files accordingly into two corpora, "FemaleBlogs.txt" and "MaleBlogs.txt." Note that our corpora are too small for answering research questions and are used for illustrative purpose only.

RUNNING THE WMATRIX ANALYSIS

Wmatrix analyses consist of uploading the data, part-of-speech tagging, semantic tagging, and a word frequency analysis:

1. Researchers log on to http://ucrel.lancs.ac.uk/wmatrix2.html with their unique username and password; free 1-month trial accounts are available for academic use.
2. The tag wizard guides users through the automatic analysis of the data (see of Figure 8.2A). We recommend switching to the "Advanced Interface" by clicking on the respective icon in the "Options" menu. The user then (a) names a work area (e.g., "FemaleBlogs"), (b) specifies a data file to be uploaded and tagged (e.g., "FemaleBlogs.txt"), and (c) starts the process by pressing "Upload now."
3. Processing the text can take between a few seconds and several minutes, depending on the complexity of the corpus. Once the analyses are finished, Wmatrix jumps to the view of the work area (e.g., "FemaleBlogs," see Figure 8.2B). The work area has links to the raw output of the word frequency, part-of-speech, and semantic tagging, as well as pull-down menus for specifying the comparison corpus.
4. The final step involves the comparison of two corpora with regard to word, part-of speech, and semantic frequencies. This can be done using one of the built-in reference corpora (e.g., BNC) or a specific comparison corpus collected by the researcher (e.g., "Male-Blogs"). Pressing "Go" runs the corpus comparison and opens up the output. The output for the word frequency (Figure 8.3A) and the semantic (Figure 8.3B) comparison of our corpora of female and male bloggers are shown in Figure 8.3.

FIGURE 8.2

A:

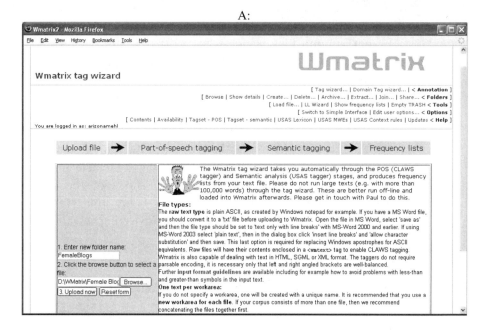

B:

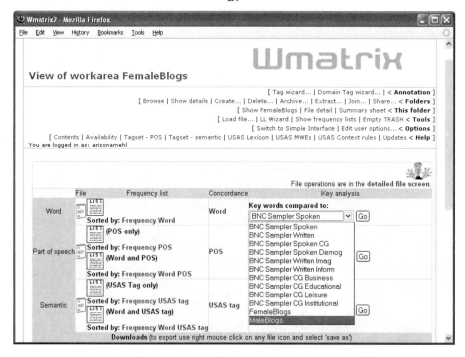

Wmatrix screenshots (advanced interface). A: Uploading the text files (Female-Blogs.txt); B: "FemaleBlogs" work area with word, part-of-speech, and semantic tag files; "MaleBlogs" is specified as the comparison corpus. From *Wmatrix* [Computer Software] by Paul Rayson, Lancaster, UK: UCREL, Lancaster University. Copyright 2000–2009 by UCREL. Reprinted with permission.

FIGURE 8.3

A.

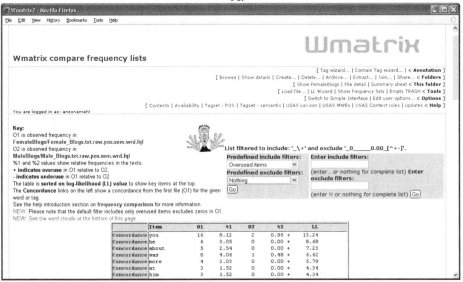

B.

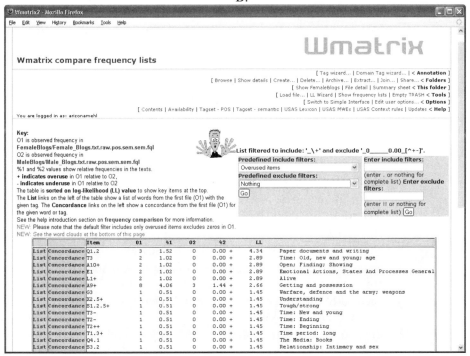

Wmatrix screenshots (advanced interface) for the "FemaleBlogs" versus "MaleBlogs" comparison. A: Output for the comparison of word frequencies; B: Output for the comparison of semantic tag frequencies. Results filtered for items overused in female blogs. From *Wmatrix* by Paul Rayson, Computer Software, Lancaster, UK: UCREL, Lancaster University. Copyright 2000–2009 by UCREL. Reprinted with permission.

INTERPRETING THE WMATRIX OUTPUT

The Wmatrix corpus comparison output provides statistical information on the overuse and underuse of individual words, part-of-speech (i.e., grammatical category), and semantic features of one text corpus relative to another. In the advanced interface (shown in the screenshots and described in our example), it supplies log-likelihood (LL) values to estimate the reliability of the between-corpora differences in words and text features; in the simple interface, it uses graphical "cloud" images to represent the most significant features. Rayson (2003) recommended 15.13 as a critical LL value ($p < .0001$) to minimize capitalization on chance due to the number of tests; however, a value of 6.63 could be used with care. It is not surprising that, given our small amount of textual data, no language differences passed this threshold. Yet, female relative to male bloggers tended to overuse the words *you, he,* and *about* (see Figure 8.3A). Consistent with these findings for single words, female bloggers also grammatically overused second person personal pronouns (*you*), third person singular subjective pronouns (*he*), and past tense verbs. Wmatrix did not reveal any semantic features that females overused greatly (all LL values < 6.63; see Figure 8.3B). Male bloggers, on the other hand, tended to overuse the word contraction '*m,* grammatically general adverbs (*really, even*), the auxiliary verb *am,* and semantically nonspecific quantifiers (*even*).

As our sample comparison of male and female blog language use reveals, Wmatrix can provide linguistic information at a very fine-grained level. The researcher's challenge then lies in conceptually interpreting the identified characteristic word-based, grammatical, or semantic group differences (Oberlander & Gill, 2006). Yet it is important to note that the extensive—and potentially overwhelming—Wmatrix output also offers unique potentials: For the first time, it is possible to automatically assess almost any grammatical feature and a wide spectrum of semantic language use features. Many of these features have escaped other, simpler ATA tools, and many of them are inherently important to behavioral scientists (e.g., use of comparatives and superlatives; personality traits; references to emotional states, health, and disease). It is because of its unique blend of computational power (automatic grammatical and semantic disambiguation), linguistic sophistication, and user friendliness that we decided to introduce it to behavioral scientists. As an easy-to-use ATA tool, Wmatrix has wide applicability and unique potentials for revealing the natural interactions among psychological and linguistic processes.

Summary and Conclusion

In this chapter, we have highlighted some of the possibilities that ATA offers for working with text-based Internet data and provided a user guide for two ATA approaches. Which of the two tools, then, should researchers use? In general, for psychologically complex (e.g., involving several continuous measures) but linguistically relatively simple phenomena (e.g., focusing on pronoun use only), a dictionary-based approach like LIWC seems optimally suited. In the contrary situation, that is, for rich linguistic data and dichotomous psychological variables, Wmatrix seems best suited. It is our conviction, though, that unique insights result from a synergy of both approaches—the joint use of the psychological LIWC categories and the linguistic Wmatrix categories (Oberlander & Gill, 2006). Additional techniques that provide a good balance between linguistic sophistication and psychological complexity include Coh-Metrix and Latent Semantic Analysis (e.g., Gill et al., 2008). More information about these tools can be found at http://cohmetrix.memphis.edu and http://lsa. colorado.edu.

Additional Resources

Chung, C. K., & Pennebaker, J. W. (2007). The psychological function of function words. In K. Fiedler (Ed.), *Social communication: Frontiers of social psychology* (pp. 343–359). New York: Psychology Press.

This chapter provides a comprehensive summary of research on psychological aspects of natural word use with a focus on variables such as gender, age, culture, personality, depression, and deception.

Mehl, M. R. (2006). Quantitative text analysis. In M. Eid & E. Diener (Eds.), *Handbook of multimethod measurement in psychology* (pp. 141–156). Washington, DC: American Psychological Association.

This chapter discusses quantitative text analysis in the context of multimethod measurement; it reviews nine text analysis strategies in psychology and classifies them on four dimensions.

Neuendorf, K. A. (2002). *The content analysis guidebook.* Thousand Oaks, CA: Sage.

This comprehensive book on content analysis as a scientific method (manual and computerized) discusses measurement issues and describes various text analysis programs. A helpful online companion is provided at http://ATAdemic.csuohio.edu/kneuendorf/content

References

Burnard, L. (Ed.). (1995). *Users' reference guide for the British National Corpus Version 1.0.* Oxford, England: Oxford University Computing Services.

Campbell, R. S., & Pennebaker, J. W. (2003). The secret life of pronouns: Flexibility in writing style and physical health. *Psychological Sciences, 14,* 60–65.

Chung, C. K., & Pennebaker, J. W. (2007). The psychological function of function words. In K. Fiedler (Ed.), *Social communication: Frontiers of social psychology* (pp. 343–359). New York: Psychology Press.

Cohn, M. A., Mehl, M. R., & Pennebaker, J. W. (2004). Linguistic indicators of psychological change after September 11, 2001. *Psychological Science, 15,* 687–693.

Forsyth, R., Ainsworth, S., Clarke, D., Brundell, P., & O'Malley, C. (2006, June). Linguistic computing methods for analysing digital records of learning. *Proceedings of the 2nd International Conference on e-Social Science,* Manchester, England.

Gill, A. J., French, R. M., Gergle, D., & Oberlander, J. (2008). The language of emotion in short blog texts. *Proceedings of the Association for Computing Machinery Conference on Computer Supported Cooperative Work (CSCW 2008)* (pp. 299–302). New York: ACM Press.

Graesser, A., McNamara, D. S., Louwerse, M., & Cai, Z. (2004). Coh-Metrix: Analysis of text on cohesion and language. *Behavioral Research Methods, Instruments, & Computers, 36,* 193–202.

Granger S., Dagneaux E., & Meunier F. (2002). *The International Corpus of Learner English* [Handbook and CD-ROM]. Louvain-la-Neuve, France: Presses Universitaires de Louvain.

Krippendorff, K. (2004). *Content analysis: An introduction to its methodology.* Beverly Hills, CA: Sage.

Landauer, T. K., Foltz, P. W., & Laham, D. (1998). An introduction to Latent Semantic Analysis. *Discourse Processes, 25,* 259–284.

Lyons, E. J., Mehl, M. R., & Pennebaker, J. W. (2006). Pro-anorexics and recovering anorexics differ in their linguistic Internet self-presentation. *Journal of Psychosomatic Research, 60,* 253–256.

McEnery, T., & Wilson, A. (1996). *Corpus linguistics.* Edinburgh, Scotland: Edinburgh University Press.

Mehl, M. R. (2006). Quantitative text analysis. In M. Eid & E. Diener (Eds.), *Handbook of multimethod measurement in psychology* (pp. 141–156). Washington, DC: American Psychological Association.

Neuendorf, K. A. (2002). *The content analysis guidebook.* Thousand Oaks, CA: Sage.

Nowson, S. (2006). *The language of weblogs: A study of genre and individual differences.* Unpublished Doctoral Dissertation, University of Edinburgh, Scotland.

Oberlander, J., & Gill, A. J. (2006). Language with character: A corpus-based study of individual differences in e-mail communication. *Discourse Processes, 42,* 239–270.

Pennebaker, J. W., Booth, R. J., & Francis, M. E. (2007). *Linguistic Inquiry and Word Count: LIWC 2007.* Austin, TX: LIWC available from http://www.liwc.net

Pennebaker, J. W., Francis, M. E., & Booth, R. J. (2001). *Linguistic Inquiry and Word Count (LIWC): LIWC 2001*. Mahwah, NJ: Erlbaum. Available from http://www.liwc.net

Pennebaker, J. W., Mehl, M. R., & Niederhoffer, K. G. (2003). Psychological aspects of natural language use: Our words, our selves. *Annual Review of Psychology, 54*, 547–577.

Popping, R. (2000). *Computer-assisted text analysis*. London: Sage.

Ramírez-Esparza, N., Pennebaker, J. W., García, A. F., & Suriá, R. (2007). La psicología del uso de las palabras: Un programa de computadora que analiza textos en Español [The psychology of word use: A computer program that analyzes texts in Spanish]. *Revista Mexicana de Psicología, 24*, 85–99.

Rayson, P. (2003). *Matrix: A statistical method and software tool for linguistic analysis through corpus comparison* (doctoral thesis). Lancaster University.

Rayson, P. (2009) *Wmatrix: A Web-based corpus processing environment*. Computing Department, Lancaster University. Available from http://ucrel.lancs.ac.uk/wmatrix/

Semin, G. R., & Fiedler, K. (1988). The cognitive functions of linguistic categories in describing persons: Social cognition and language. *Journal of Personality and Social Psychology, 54*, 558–568.

Shapiro, G., & Markoff, J. (1997). A matter of definition. In C. W. Roberts (Ed.), *Text analysis for the social sciences: Methods for drawing statistical inferences from texts and transcripts* (pp. 8–31). Mahwah, NJ: Erlbaum.

Stone, L. D., & Pennebaker, J. W. (2002). Trauma in real time: Talking and avoiding online conversations about the death of Princess Diana. *Basic and Applied Social Psychology, 24*, 172–182.

West, M. D. (Ed.). (2001). *Theory, method, and practice in computer content analysis*. New York: Ablex.

Wilson, A., & Moudraia, O. (2006). Quantitative or qualitative content analysis? Experiences from a cross-cultural comparison of female students' attitudes to shoe fashions in Germany, Poland, and Russia. In A. Wilson, P. Rayson, & D. Archer (Eds.), *Corpus linguistics around the world* (pp. 203–217). Amsterdam: Rodopi.

Wolf, M., Horn, A. B., Mehl, M. R., Haug, S., Kordy, H., & Pennebaker, J. W. (2008). Computergestützte quantitative Textanalyse: Äquivalenz und Robustheit der deutschen Version des Linguistic Inquiry and Word Count [Computerized quantitative text analysis: Equivalence and robustness of the German adaptation of Linguistic Inquiry and Word Count]. *Diagnostica, 54*, 85–98.

Zijlstra, H., van Meerveld, T., van Middendorp, H., Pennebaker, J. W., & Geenen, R. (2004). De Nederlandse versie van de "Linguistic Inquiry and Word Count" (LIWC): Een gecomputeriseerd tekstanalyseprogramma. [Dutch version of Linguistic Inquiry and Word Count (LIWC), a computerized text analysis program]. *Gedrag & Gezondheid, 32*, 271–272.

TRANSPORTING TRADITIONAL METHODOLOGIES TO THE WEB

IV

Ulrich Schroeders, Oliver Wilhelm, and Stefan Schipolowski

Internet-Based Ability Testing

9

C ompared with traditional paper-and-pencil-tests (PPTs), Internet-based ability testing (IBAT) seems to offer a plethora of advantages—cost-effective data collection 24/7 from all over the world, enriching static content by implementing audio and video, registering auxiliary data such as reaction times, and storing data at some central location in a digital format—all this seems like the fulfillment of a researcher's dream. However, despite these auspicious possibilities, the dream can rapidly dissolve if the accompanying constraints and limitations of testing via the Internet are neglected.

This chapter focuses on ways to effectively use the Internet for ability testing and on procedures that might help overcome the inherent shortcomings of the medium. Wherever possible, we highlight how to minimize the adverse effects and provide practical advice to improve Web-based measures. We offer a detailed step-by-step guide that helps structure the entire testing process—from the definition of the construct being assessed through the presentation of the findings. At the end of each step, we present an example of our own ongoing research that illustrates how to implement the step. Specifically, we show that IBAT can be used as a method to quickly gather information about characteristics of a measure in an early stage of development.

We hope to provide information that helps you to decide which kinds of ability tests can be administered via the Internet and for which purposes, and we give practical advice on how to implement your own test on the Internet. For an easy start, the provided code can be easily altered to meet specific needs. We also list links to detailed information on PHP, a powerful programming language enabling researchers to use the whole functionality of a PC, for example, video streaming (see Additional Resources A, C).

Drawbacks of IBAT might be a lack of experimental control with regard to individual differences in motivation and time allocation, respectively, or inscrutable distortions due to self-selection processes. Also, programming an online measure can be quite time consuming. But once you overcome or know how to tackle these obstacles, IBAT is a powerful tool for collecting data independent of time and geographic constraints.

Implementation of the Method

We provide a step-by-step-guide consisting of 10 interlocked steps to implement ability tests on the Internet. First, we describe each step conceptually. Then, we highlight how to implement the step in practical terms and provide a research example in a gray-shaded box illustrating how the method was used in order to advance test construction. The whole implementation process comprises the following steps:

1. Describe the purpose of the study, define and operationalize the construct to be assessed, and generate the items.
2. Design the presentation of items carefully.
3. Decide how to record the response.
4. Define how to score the input and give feedback.
5. Choose how to select items.
6. Write a clear instruction and debriefing.
7. Thoroughly check the content both offline and online.
8. Publish and advertise the ability test on the Web.
9. Manage and process the collected data.
10. Analyze the data; interpret and communicate the results.

Necessarily, our discussion of these points is just a short sketch and not an exhaustive collection of issues that ought to be considered in IBAT. Contingent on specific research questions, additional issues might arise at each step.

STEP 1. DESCRIBE THE PURPOSE OF THE STUDY, DEFINE AND OPERATIONALIZE THE CONSTRUCT TO BE ASSESSED, AND GENERATE THE ITEMS

A clear definition of the purpose and subject of measurement is critical for any ability testing. This task is primarily psychological and not technological. Prior to planning an experiment, you have to answer some general questions, such as

- What is the purpose of the study?
- What is the construct I want to measure? and
- What prior knowledge and measurement expertise is available for this construct?

We intended to develop a measure assessing individual differences in the figural facet of fluid intelligence for student selection purposes that is in principle amenable to automated item generation. We decided to use matrices as item material because it has been shown to be a prototypical task to measure fluid intelligence (Cattell, 1963; Horn & Cattell, 1966). In order to advance test construction we conducted an Internet study to shed some light on the following research questions:

1. Are item difficulties distributed in a way that allows precise measurement in a broad ability range?
2. Can certain item attributes predict item difficulty to a substantial degree? and
3. Can all items of the test be regressed on a single latent factor in a confirmatory measurement model with adequate fit?

Any ability testing procedure can be regarded as running through a fixed recursive cycle (see Figure 9.1) that has been conceptualized in the architecture for assessment delivery systems, part of the evidence-centered design (Mislevy & Riconscente, 2006; Mislevy, Steinberg, Almond, & Lukas, 2006). The cycle comprises four processes—presenting an item, processing and scoring the response, and selecting the next item—which can be found in Steps 2 through 5 of our implementation process.

STEP 2. DESIGN THE PRESENTATION OF ITEMS CAREFULLY

The static content of a PPT can be enriched in many ways by including audio, video, or interactive flash objects (for a good example, see Richardson et al., 2002; for a detailed overview, see chap. 4, this volume). The possibility to use an interactive item format that comes closer to real-life settings might even be associated with an increase in external validity (Kyllonen & Lee, 2005; Melnick & Clauser, 2006). However, using

FIGURE 9.1

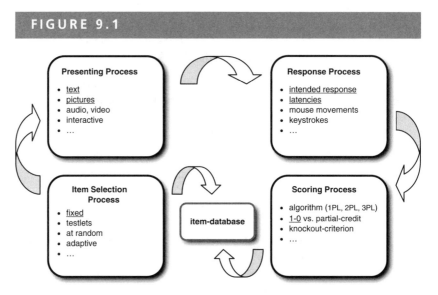

The testing cycle in dependence on the four-process architecture for assessment delivery systems. Underlined bullet points symbolize the instances used in our example.

multimedia in testing can also turn into a disadvantage: Developing audio or video items is a time-consuming endeavor for test developers and might distract test takers from completing the task. We suggest that the use of multimedia in IBAT depends on the answer to the question, is the new content affecting the measurement process positively?

Aside from such conceptual considerations, the presentation process is influenced by both software aspects (e.g., layout, browser functionality) and hardware aspects (e.g., Internet connection, display resolution). Much research has been conducted concerning technological issues, for example, the influence of font characteristics or the number of lines on the legibility of online texts (for a summary, see Leeson, 2006). The available hardware that also constrains the presentation process is improving expeditiously, making it hard to give long-lasting advice. At present, online tests should be designed to be compatible with different browsers and operating systems, with a resolution of at least 800×600 pixels, and texts should be in TrueType fonts. Useful supplementary information is provided in the *International Guidelines on Computer-Based and Internet Delivered Testing* (International Test Commission, 2005).

Throughout the test, we abstained from multimedia features and reduced the test material to basic text and graphics. By preferring a simply structured PHP and MySQL solution (see Additional Resources C, D, E) instead of a resource-demanding Java application, we attempted to overcome limiting factors such as hardware resources or browser settings. By editing the MySQL database, you can easily modify the source code to suit your needs. (see Figure 9.2)

STEP 3. DECIDE HOW TO RECORD THE RESPONSE

The possibilities of recording auxiliary data in IBAT are manifold. For instance, one can easily retrieve response latencies, IP addresses, or the resolution of the user's display. Data on time aspects can be a valuable tool to scan the data and exclude responses with extremely low or high latencies, as these are probably invalid observations. IP addresses often can be used to derive information on the test taker's geographic location. However, collecting more data, such as keystrokes or mouse pointer movements, does not necessarily imply obtaining more valuable information. For example, relatively little is known about correct decision

FIGURE 9.2

```php
// presentation of the items incl. the example ($intro = 0 & $page = 1)
// as long as the $page variable does not exceed the number of items
   if ($page <= $numberOfItems) {
     echo "<td rowspan='7' height='360' width='310'><br>";
     if ($intro == 1)
       echo "<strong>item ".$page." /".$numberOfItems."</strong>";
     else echo "<strong>Example</strong>";

// the matrices are labeled item1.gif, item2.gif, etc. and are
// stored in the folder images
     echo "<img src='images/item".$page.".gif' alt='item ".$page."'
          width='300' height='300' /></td></tr>";
     echo "<tr align='center'>";

// the response alternatives are image files labeled item1response.gif,
// item2response.gif, etc. The chosen response alternative is determined
// depending on the image area where participants click.
     echo "<td><input type='image' src='images/item".$page."response.gif'
          name='item".$page."' width='300' height='300' border='0'
          onFocus='this.blur()';>
          <div style='font-size:9pt; font-style:italic;'>
          Click on the correct response</div></td></tr>";
   }
```

Source code for presenting an item; for notes, see lines that are commented out (marked with "//" and grayed out).

speed in measuring fluid or crystallized intelligence (Danthiir, Wilhelm, & Schacht, 2005). The critical issue is whether such incidental data will add valuable information to the measurement. In the case fluid and crystallized intelligence measuring reaction times will not pay off if the test instruction did not promise credit for fast correct responses. However, for other constructs, such as mental speed, recording both accuracies and response latencies is essential for an adequate assessment of performance.

If it is necessary to record response times very accurately, say in the range of milliseconds—for example, when assessing small effects such as negative priming—we currently recommend implementing Java applications with Web Start technology (see Additional Resources G). In this context, it is important to distinguish the precision of estimates for the mean of a distribution from the precision of estimates of person parameters. In ability measurement, we are usually predominantly concerned with the precision of person parameters.

In our test, we registered both responses and latencies (see Figure 9.3), the former being the behavior in question, the latter for the purpose of getting indicators of irregular response behavior (extremely low or high latencies) to exclude such supposedly invalid observations from subsequent analyses. For the present purpose, accuracy of time measurement in the range of seconds was sufficient. All data were stored in a central MySQL database (see Additional Resource D).

FIGURE 9.3

```
// function to determine the reaction time
   function getmicrotime() {
     list($usec, $sec) = explode(" ",microtime());
     return ((float)$usec + (float)$sec);
   }

...
// the variable $_GET[timeStart]) is transferred using GET method
   if(isset($_GET[timeStart])) {
     $timeEnd = getmicrotime();
     $time = $timeEnd - $_GET[timeStart];
   }
```

Source code to calculate the reaction times.

STEP 4. DEFINE HOW TO SCORE THE INPUT AND GIVE FEEDBACK

Through IBAT, the test taker gets the opportunity to receive automatically generated feedback right after completing a test. The report can include

- a visualization of the testee's results, for example, a distribution of the scores in the sample with a marker indicating the subject's position;
- a template-based text that summarizes the testee's results; and
- the correct answers in comparison to the given answers with an explanation (e.g., on how to derive the correct conclusion in case of a deductive reasoning test).

Bartram (1995) noted that most computer-generated test reports are designed for the test administrator rather than the test taker. Therefore, one should carefully consider which data to report in which way. Researchers should conceive the feedback for the testees not as a compulsive "must" but as an opportunity to communicate study results and to win participants for upcoming studies by providing encouraging and interesting results. Feedback is not a one-way street. Test designers can benefit from asking participants for their feedback and opinion. A short survey can contain questions such as

- Did you encounter any technical problems (e.g., layout, database connectivity)?
- Did you miss any vital information in the instruction or the test material?
- Was the feedback encouraging, interesting, transparent, explicit, and concrete?
- Do you have additional comments concerning the testing procedure and content?

In our study the answers were scored dichotomously (*incorrect–correct;* see Figure 9.4). The sum of correct responses was reported to the participants immediately after finishing the test. Later, a more sophisticated report about the subject's rank position and general results of the study could be retrieved by test takers using a Web interface.

STEP 5. CHOOSE HOW TO SELECT ITEMS

This step attends to the mechanism of item selection. As with computer-based test administration, Web-based ability testing enables researchers to score item responses directly by accessing an underlying item database and thus selecting the next item. Particularly in adaptive testing, it is worthwhile to implement online scoring algorithms. A sophisticated

FIGURE 9.4

```
// The chosen response alternative ($itemInput) is compared to the
// corresponding solution ($solution[$sol]); if identical the input is
// scored 1 ($itemInputCorrect = 1) else the answer is incorrect
// ($itemInputCorrect = 0;)
   ...
  $solution = array(6, 9, 8);
  for ($l=0; $l<=$numberOfItems; $l++) {
    if($page == $l+1 && $itemInput == $solution[$l]) {
      $itemInputCorrect = 1;
      $correctTotal++;
      break;
    }
    else $itemInputCorrect = 0;
  }
```

Source code for scoring an item; for notes, see lines that are commented out (marked with "//" and grayed out).

scoring algorithm is of interest not only for the selection of the next item but also to detect cheating or unwanted item exposure prior to testing. Indicators of cheating can be found by analyzing and detecting unexpected response patterns, that is, a correct response on a difficult item in spite of incorrect answers on easier items or response time anomalies (van der Linden & van Krimpen-Stoop, 2003). Furthermore, the item estimates in the database can incrementally become more precise with data from additional participants being reported, thus enabling more precise estimations of the person's ability.

The development of item response theory has led to computerized adaptive testing (CAT; van der Linden & Glas, 2000), abandoning the rigid item order and enabling response-dependent testing that optimizes information gain per item. A variety of procedures has been developed in the past addressing specificities of CAT such as test length and stopping rules (Gershon, 2005). However, findings on the equivalence of fixed versus adaptive item presentation are inconsistent: Whereas some authors found no effect (Mason, Patry, & Bernstein, 2001; Mead & Drasgow, 1993), other studies show substantial differences (Kim, 1999; Wang, Jiao, Young, Brooks, & Olson, 2007, 2008). In practice, the algorithm of item selection is influenced by issues such as

- the scope of the item database,
- the information available on the item difficulties, and
- the time for solving an item in relation to time demands for the whole test.

In the early stage of test development, we simply used the PC as an electronic page turner, transferring test material from paper to a digital format in a fixed order. This design was sufficient for the purpose of the pilot study, namely, collecting initial data needed to revise an early version of the test.

STEP 6. WRITE A CLEAR INSTRUCTION AND DEBRIEFING

The sixth step deals with formulating a clear instruction and a debriefing to eliminate potential drawbacks such as cheating or individual differences in test takers' motivation. Cheating is one of the major problems of unproctored ability testing through the Web that can hardly be ruled out completely. However, the extent and probability of cheating can be diminished in different ways. Participants' compliance might be improved or controlled by emphasizing certain aspects in the instruction:

- guarantee anonymity of data if not contraindicated;
- stress the importance of not cheating for receiving valuable feedback;
- do not offer financial gratification or other rewards for good test results, as this is likely to foster cheating (e.g., high incentives for participation raise the number of subjects participating more than once; Frick, Bächtiger, & Reips, 1999); and
- appeal to the honesty of test takers when debriefing and explicitly ask test takers whether disallowed aids were used.

Apart from cheating, another reason for reduced external validity in IBAT may be a different motivation for test taking in a traditional versus an Internet group: Compared with the purposes of PPT, in which examinees frequently take a test for obvious reasons (e.g., job application, counseling), the purposes of IBAT are more varied, and in many instances examinees might participate out of curiosity or other less obvious reasons (Milbradt, 2008). Intrinsic motivation for low-stakes testing is likely to be higher for IBAT because subjects are self-selected and apparently willing to invest time and effort in participating (Wilhelm & McKnight, 2002). Due to this self-selection process and selection resulting from restricted access to the Internet, test data gained through IBAT tend to have slightly higher mean scores than traditionally recorded PPT data with samples unrestricted in range (Wilhelm, Witthöft, & Größler, 1999). A general test-taking questionnaire (Arvey, Strickland, Drauden, & Martin, 1990) can be applied to take into account the reasons for participating in Web-based ability studies as a correlate of differences in means and covariances in manifest and latent variables. In their framework, Arvey et al. distinguished nine facets of test taking motivation, including lack of concentration, comparative anxiety, general need for achievement, and consequences of performance. It can be useful to implement self-assessment surveys to analyze the effects of motivation on the assessment results and, if justified, exclude testees with very low motivation (Musch & Klauer, 2002). Other techniques, such as the high-hurdle technique or warm-up strategy, have also been proposed (Reips, 2002). We suggest the following course of action to minimize the influence of individual differences in motivation:

- recruit test takers via reliable sources;
- offer explicit information about the test's purpose and expenditure of time prior to testing to prevent dropout of participants (nevertheless, data from prematurely terminated tests are more useful than no data at all, because they allow inferences about predictors of dropout);
- to avoid misunderstandings of the instruction, it is advisable to offer some trial items for the participants familiarizing the participants with the items and teaching them how to respond (see Figure 9.5; in addition to the correct answer, the way of deriving the

FIGURE 9.5

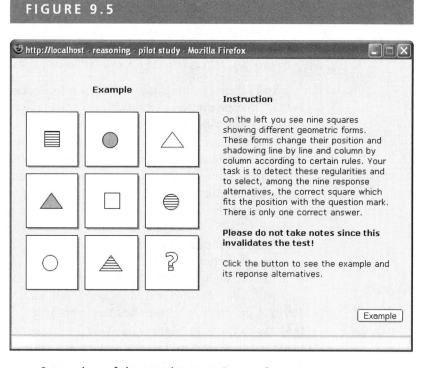

Screenshot of the matrices test: Instruction.

solution should be explained to ensure that all participants fully understand what they are supposed to do).

STEP 7. THOROUGHLY CHECK THE CONTENT BOTH OFFLINE AND ONLINE

Prior to publishing on the Web and approaching a large number of potential participants, you should thoroughly test the measurement instrument, simulating various unintended reactions (e.g., going back in the sites' history, having extreme latencies) and different browser settings (e.g., disabling JavaScript or cookies, switching to a minimal resolution of 800×600).

To check layout matters, you can use one of several special open-source online services (see Additional Resource H) that have been developed to offer screenshots of your Web site for dozens of browsers. Careful test runs are crucial for collecting complete and high-quality data. If you want to shortcut the cycle of testing the instrument, editing the source code, copying the files to a server, and checking the changes again, you can use the open-source product XAMPP, an all-in-one package consisting of the Apache HTTP Server with MySQL connectivity and PHP support (see Additional Resource B). It allows you to test the functionality of static and dynamic Web sites offline without uploading the content to a server.

> We checked whether all content was displayed on the screen without scroll bars, even with a minimal resolution, and whether the layout and functionality were invariant over different browsers and settings. The storage of the input was checked offline using XAMPP. To prevent participants from skipping items or the demographic questionnaire, we used a JavaScript fragment that enables the "Proceed" button only after valid information has been entered.

STEP 8. PUBLISH AND ADVERTISE THE ABILITY TEST ON THE WEB

Researchers can advertise their ability tests to recruit participants (see Web Sites for Recruiting Participants in the Additional Resources). In the context of promoting a measure, the issue of external validity of ability test data gathered through the Internet deserves some attention. Approximately 578 million people in Asia, 384 million people in Europe, and 248 million people in North America have access to the Internet (Internetworldstats, 2008). According to the statistics, the group of Internet users is not representative of the population. Given that a high proportion of psychological research is based on student samples highly restricted in their range of age, ability, and many sociodemographic characteristics, the limitations of Internet samples are probably overrated. In fact, by using the Internet, researchers can collect data from specific samples hardly available through other media. For example, geographically distant samples such as mathematics students in Iceland or geographically distributed samples such as support groups for people affected by some rare disease might become accessible. In most cases, it will be desirable to collect some biographical data on sex, age, and educational background to test for critical group differences (Wicherts, 2007), adverse impact, or bias (American Educational Research Association, American Psychological Association, & National Council on Measurement in Education, 2002). To account for recruitment effects and to optimize your recruitment process, it is advisable to ask participants about the source that called their attention to the study.

> The purpose of our newly developed test was to measure an aspect of fluid intelligence in the context of college admission into German psychology programs. Thus, the broadly defined population in question was that of recent high school graduates preparing for college. Therefore, apart from advertising on our home page and some German sites containing a collection of psychological tests, we also addressed a mailing list of all psychology students of our university. To check for group differences between psychology students and high school graduates, we asked subjects about their highest educational degree and whether they envisage studying psychology.

STEP 9. MANAGE AND PROCESS THE COLLECTED DATA

The data stored in the MySQL database can easily be exported in .xls or .csv format using phpMyAdmin (see Additional Resource F). After exporting the data set, it is essential to inspect all data with incomplete, aberrant, or deviant answer patterns to get more reliable and valid results. It is very hard to give universal advice on this issue, but this list should give you some guidance:

- Examine the cases with missing data and exclude cases with unacceptably high proportions of missing information in the response vector if you assume that such cases predominantly contribute invalid information to your data;
- consider how many seconds it takes to provide an answer that is not given completely at random, and then, on the basis of these considerations, predefine a threshold below which no valid conclusion can be derived from the remaining data points;

- depending on the assessed ability, you can also define a maximum answer time such that above this threshold, some circumstances must have occurred (e.g., looking up the correct answer on the Internet, breakdown of the Internet connection), supposedly invalidating the data point;
- graphically display your data to detect uni-, bi-, and multivariate outliers;
- try to come up with implausible data constellations (e.g., a 12-year-old boy working as stuntman, holding a PhD degree) and exclude such observations as probably invalid;
- make your rules for exclusion transparent and justify them psychologically; and
- compare excluded and included observations on available and supposedly valid information to learn more about the reasons for premature termination.

We collected data from 447 subjects who started with the test on the instruction page. Due to incomplete data, 103 cases were excluded. As a next check, observations with response times shorter than 5 seconds or longer than 20 minutes were considered invalid and treated as incorrect. As a corollary, 9 additional participants who had less than 50% valid item responses were excluded. Hence, 335 subjects were included in the following analyses, with 33 subjects having one or more missing data points.

STEP 10. ANALYZE THE DATA; INTERPRET AND COMMUNICATE THE RESULTS

As a last step in the testing cycle, the data are analyzed. The main purpose of the analysis is to answer the questions formulated at the beginning of the testing cycle. In completing this step, it is critical to apply scientific know-how that is tailored to the problem. It might be tempting to apply convenient but suboptimal data analytic techniques, but in the long run a more profound data analysis is essential for the interpretation of results. We therefore encourage you to, as the saying goes, "just say no" to using default options in factor analysis, to falling back on unsatisfactory reliability estimates, to using sum scores without checking critical assumptions, and to not worrying about the relations of a measure with related and unrelated criteria. A profound understanding of validity issues (Borsboom, 2006), test theoretical considerations (McDonald, 1999; e.g., the differences between exploratory factor analysis vs. principal-components analysis discussed in Preacher & MacCallum, 2003), and the application of powerful software (see Byrne, 2001; L. K. Muthén & Muthén, 2006) should be a standard if you use ability measures for research purposes. After analyzing your data critically and interpreting the results with respect to your initial hypotheses, it is important to relate the findings to substantial theories—after all, the testing was done to address a psychological problem.

> On the basis of the data we collected, we can answer our research questions as follows:
>
> *ad 1: Distribution of item difficulties.* The mean difficulty of .71 is above the discriminating optimum of .50, indicating that the test was relatively easy. The 17 items cover a satisfactory range of item difficulties from .41 to .94.
>
> *ad 2: Prediction of item difficulties.* Our aim was to predict item difficulties by the number of elements and rules involved in constructing the stimulus material. Thus, the research question was to what extent the data supported the rational of the test construction principles applied. The correlation between item difficulty and number of elements and rules is $r = -.67$ ($p < .01$), indicating that items with more elements and rules are more difficult. This finding lends some support to the construction rationale.
>
> *ad 3: Psychometric analyses.* The part–whole corrected biserial item–total correlations with a mean of $r_{it} = .59$ and the internal consistency on the basis of a tetrachoric correlation matrix (Nunnally & Bernstein, 1994) are very good (Cronbach's $\alpha = .92$). Confirmatory factor analysis (weighted least squares mean and variance adjusted estimator; B. O. Muthén, 1984) supports a single-factor model that fits the data very well: $\chi^2(68, N = 335) = 72.7$, $p = .33$, comparative fit index = .996, root-mean-square error of approximation = .014. Therefore, the hypothesis that all 17 items measure a single latent variable is confirmed. The reliability of this latent factor was $\omega = .92$.

Conclusion

In this chapter, we illustrated ability testing via the Internet by going through a step-by-step guide. The example we used fosters test development, but other research purposes are also possible. Usually, test construction runs through all steps of the testing cycle more than once to optimize a measure. We pointed out some limitations of IBAT that can cause serious drawbacks, accompanied by some recommendations on how to minimize them. If the pitfalls can be avoided, the Internet proves to be a productive tool for administering ability tests. Aside from the simplicity and economy of testing, the Internet can also be helpful in recruiting a more heterogeneous sample than the often-tested sample of psychology students.

Additional Resources

PROGRAMMING

A. http://www.apa.org/books/resources/gosling

Here you can find the source code of the examples used in this chapter. Additional explanations of the programming are given in the source code directly.

B. http://www.apachefriends.org/en/xampp.html

The Apache HTTP Server with PHP and MySQL is an easy way to test the functionality and design of your test offline.

C. http://www.phpfreaks.com/

This site offers a great collection of articles about PHP programming, a code library, tutorials, and forums where you can post questions.

D. http://dev.mysql.com/doc/refman/6.0/en/index.html

You can use the MySQL 6.0 reference manual (located here) to look up the MySQL syntax, for instance, on how to write different types of data to your database.

E. http://www.php.net/docs.php

Here you can find the official PHP documentation.

F. http://www.phpmyadmin.net

On this site you can get additional information about the program phpMyAdmin that is an easy to use tool for creating, editing, and exporting MySQL databases.

G. http://java.sun.com/products/javawebstart

On Java's official site you can inform yourself about the Java Web Start technology if you are interested in accurately recording reaction times.

H. http://browsershots.org/

Use this link to test the layout of your Web site in different browsers.

WEB SITES FOR RECRUITING PARTICIPANTS

- http://www.queendom.com
- http://www.psychtests.com
- http://psych.hanover.edu/Research/exponnet.html

References

American Educational Research Association, American Psychological Association, & National Council on Measurement in Education. (2002). *Standards for educational and psychological testing.* Washington, DC: American Educational Research Association.

Arvey, R., Strickland, W., Drauden, G., & Martin, C. (1990). Motivational components of test taking. *Personnel Psychology, 54*, 695–716.

Bartram, D. (1995). The role of computer-based test interpretation (CBTI) in occupational assessment. *International Journal of Selection and Assessment, 3*, 31–69.

Borsboom, D. (2006). The attack of the psychometricians. *Psychometrika, 71*, 425–440.

Byrne, B. M. (2001). *Structural equation modeling with AMOS: Basic concepts, applications, and programming.* Hillsdale, NJ: Erlbaum.

Cattell, R. B. (1963). Theory of fluid and crystallized intelligence: A critical experiment. *Journal of Educational Psychology, 54*, 1–22.

Danthiir, V., Wilhelm, O., & Schacht, A. (2005). Decision speed in intelligence tasks: Correctly an ability? *Psychology Science, 47*, 200–229.

Frick, A., Bächtiger, M. T., & Reips, U.-D. (1999). Financial incentives, personal information, and drop-out rate in online studies. In U.-D. Reips, B. Batinic, W. Bandilla, M. Bosnjak, L. Gräf, K. Moser, & A. Werner (Eds.), *Current Internet science—Trends, techniques, results.* Retrieved January 12, 2008, from http://gor.de/gor99/tband99/pdfs/a_h/frick.pdf

Gershon, R. C. (2005). Computer adaptive testing. *Journal of Applied Measurement, 6*, 109–127.

Horn, J. L., & Cattell, R. B. (1966). Refinement and test of the theory of fluid and crystallized intelligence. *Journal of Educational Psychology, 57*, 253–270.

International Test Commission. (2005). *International guidelines on computer-based and Internet delivered testing.* Retrieved January 12, 2008, from http://www.intestcom.org/guidelines/index.php

Internetworldstats. (2008). *Internet usage statistics.* Retrieved January 12, 2008, from http://www.internetworldstats.com/stats.htm

Kyllonen, P. C., & Lees, S. (2005). Assessing Problem Solving in Context. In O. Wilhelm & R. W. Engle (Eds.), *Handbook of understanding and measuring intelligence.* (pp. 11–25). Thousand Oaks: Sage Publications.

Kim, J.-P. (1999, October). *Meta-analysis of equivalence of computerized and P&P tests on ability measures.* Paper presented at the annual meeting of the Mid-Western Educational Research Association, Chicago.

Leeson, H. V. (2006). The mode effect: A literature review of human and technological issues in computerized testing. *International Journal of Testing, 6*, 1–24.

Mason, B. J., Patry, M., & Bernstein, D. J. (2001). An examination of the equivalence between non-adaptive computer-based and traditional testing. *Journal of Educational Computing Research, 24,* 29–39.

McDonald, R. P. (1999). *Test theory: A unified treatment.* Mahwah, NJ: Erlbaum.

Mead, A. D., & Drasgow, F. (1993). Equivalence of computerized and paper-and-pencil cognitive ability tests: A meta-analysis. *Psychological Bulletin, 114,* 449–458.

Melnick, D. E., & Clauser, B. E. (2006). Computer-based testing for professional licensing and certification of health professionals. In D. Bartram & R. K. Hambleton (Eds.), *Computer-based testing and the Internet: Issues and advances* (pp. 163–186). New York: Wiley.

Milbradt, A. (2008). Quality criteria in open source software for computer-based assessment. In F. Scheuermann & A. G. Pereira (Eds.), *Towards a research agenda in computer-based assessment: Challenges and needs for European educational measurement.* Retrieved January 12, 2008, from http://crell.jrc.it/CBA/EU-Report-CBA.pdf.

Mislevy, R. J., & Riconscente, M. M. (2006). Evidence-centered assessment design. In S. M. Downing & T. M. Haladyna (Eds.), *Handbook of test development* (pp. 61–90). Mahwah, NJ: Erlbaum.

Mislevy, R. J., Steinberg, L. S., Almond, R. G., & Lukas, J. F. (2006). Concepts, terminology, and basic models of evidence-centered design. In D. M. Williamson, R. J. Mislevey, & I. I. Bejar (Eds.), *Automated scoring of complex tasks in computer-based testing* (pp. 15–47). Hillsdale, NJ: Erlbaum.

Musch, J., & Klauer, K. C. (2002). Psychological experimenting on the World Wide Web: Investigating content effects in syllogistic reasoning. In B. Batinic, U.-D. Reips, & M. Bosnjak (Eds.), *Online social sciences* (pp. 181–212). Seattle, WA: Hogrefe & Huber.

Muthén, B. O. (1984). A general structural equation model with dichotomous, ordered categorical, and continuous latent variable indicators. *Psychometrika, 49,* 115–132.

Muthén, L. K., & Muthén, B. O. (2006). *Mplus.* Los Angeles: Muthén and Muthén.

Nunnally, J. C., & Bernstein, I. H. (1994). *Psychometric theory* (3rd ed.). New York: McGraw-Hill.

Preacher, K. J., & MacCallum, R. C. (2003). Repairing Tom Swift's electric factor analysis machine. *Understanding Statistics, 2,* 13–43.

Reips, U.-D. (2002). Standards for Internet-based experimenting. *Experimental Psychology, 49,* 243–256.

Richardson, M., Baird, J.-A., Ridgway, J., Ripley, M., Shorrocks-Taylor, D., & Swan, M. (2002). Challenging minds? Students' perceptions of computer-based World Class Tests of problem solving. *Computers in Human Behavior, 18,* 633–649.

van der Linden, W. J., & Glas, G. A. W. (2000). *Computerized adaptive testing: Theory and practice.* Dordrecht: Kluwer Academic Publishers.

van der Linden, W. J., & van Krimpen-Stoop, E. M. L. A. (2003). Using response times to detect aberrant responses in computerized adaptive testing. *Psychometrika, 68,* 251–265.

Wang, S., Jiao, H., Young, M. J., Brooks, T., & Olson, J. (2007). A meta-analysis of testing mode effects in grade K–12 mathematics tests. *Educational and Psychological Measurement, 67,* 219–238.

Wang, S., Jiao, H., Young, M. J., Brooks, T., & Olson, J. (2008). Comparability of computer-based and paper-and-pencil testing in K–12 reading assessments: A meta-analysis of testing mode effects. *Educational and Psychological Measurement, 68,* 5–24.

Wicherts, J. M. (2007). *Group differences in intelligence test performance.* (Doctoral thesis) Universiteit van Amsterdam, Amsterdam.

Wilhelm, O., & McKnight, P. E. (2002). Ability and achievement testing on the World Wide Web. In B. Batinic, U.-D. Reips, & M. Bosnjak (Eds.), *Online social sciences* (pp. 151–180). Seattle, WA: Hogrefe & Huber Publishers.

Wilhelm, O., Witthöft, M., & Größler, A. (1999). Comparisons of paper-and-pencil and Internet administrated ability and achievement test. In P. Marquet, A. Mathey, A. Jaillet. & E. Nissen (Eds.), *Proceedings of IN-TELE 98* (pp. 439–449). Berlin: Peter Lang Publishing.

John A. Johnson

Web-Based Self-Report Personality Scales

10

T his chapter describes how to build Web pages to administer, score, and provide feedback for self-report scales. Included in this chapter are samples of HTML and Perl code for displaying scale items on the Web, prompting participants who skip items, instantly checking for nonresponsiveness (responding inconsistently, randomly, or with strings of the same response category without regard to content), storing item responses for later analyses, computing scale scores, and producing feedback in the form of a narrative report. Shorter examples of code appear directly in this chapter; longer examples can be accessed at the supplementary Web site for this chapter as described in the Additional Resources section at the end of the chapter.

Self-reports sum a person's responses over a set of items, resulting in a single aggregated score that measures the degree or level of a single psychological state or trait. Although self-reports are often referred to as "questionnaires," they should not be confused with survey or opinion questionnaires (see chap. 12, this volume), which treat responses to each item separately rather than in aggregates. Self-reports are also often called "tests" but should not be confused with ability or aptitude tests (see chap. 9, this volume), whose items are problems to be solved rather than statements with which participants judge applicability or agreement.

Implementing a Self-Report Measure on the Web

Anyone who possesses the knowledge of Web page authoring described in chapters 2 through 6 of Fraley (2004) can create Web-based self-report scales through the following steps:

- Consider what you want to accomplish with Web-based assessment;
- choose (a) self-report scale(s) that will allow you to accomplish your goals;
- decide on recruiting strategies for obtaining participants;
- design the front page that invites participants to complete the scale;
- write and test files that present the items to participants;
- write and test script files for error checking, if desired;
- write and test script files that score the scale and record the results; and
- construct a routine for providing narrative feedback, if desired.

Each stage involves choosing from options according to one's purposes. Choices at each stage impact later stages. Next, I describe some of these options at each stage and how to implement them.

WHAT DO YOU WANT TO ACCOMPLISH?

Being clear about what you want to accomplish helps you to make good decisions in each of the remaining steps of Web-based measurement. One simple research goal for self-reports is to ascertain a single measure's psychometric properties, such as item endorsement frequencies, distribution of scores, scale reliability, and factor structure. Additional simple research goals include examining the scale's relation to demographic variables, such as age, sex, and ethnicity. If your goals fall into the category of simple research, your self-report scale is short, and you are not gathering a lot of other information about participants, you can present all of your items on a single Web page. However, if your goals are more complex (e.g. assessing construct validity by examining relations among several constructs with a longer, multiscale inventory or several different self-report scales), you will want to write scripts that spread the items across multiple Web pages. If you wish to have participants return to your site at a later time to retake your measure or to complete other measures, or if you are gathering information on the participants from informants (see chap. 11, this volume) or other sources, you will need to create an identifier for participants that allows you to match scores from the same participant while protecting his or her privacy.

Your goals will therefore determine the number of Web pages you will need to construct and what kind of identifying information to record. Your goals will also determine your recruiting strategy for inviting participants to complete your Web-based measure.

CHOOSE SELF-REPORT SCALES

Other than the standard concerns about conceptual relevance, reliability, and validity, the major consideration in choosing self-report scales is that they can be placed on the Web without violating copyright laws. Commercial test publishers rarely grant permission to reproduce their scales on the Web; authors of noncommercial measures are far more likely to do so. Researchers who wish to avoid altogether the bother of obtaining permission to place scales on the Web will find hundreds of scales in the public domain at the International Personality Item Pool (IPIP) Web site, http://ipip.ori.org (Goldberg, et al., 2006).

DECIDE ON RECRUITING STRATEGIES

Options for inviting participants fall on a continuum of active versus passive recruiting. Actively pursuing participants is useful when you are interested in collecting a modest amount of data in a relatively short period of time. A more passive approach involves waiting for Web surfers to discover your site and allowing knowledge of the site to spread by word of mouth. Researchers who are not in a hurry can recruit hundreds of thousands of participants in a few years' time with this approach. A mixed strategy involves recruiting a certain number of participants at the outset and then leaving the site up for additional people to add their responses to the data pool.

If you wish to actively recruit individuals with particular traits or interests (e.g., shy people; people who are passionate about politics), you can announce your research study in relevant Internet communities. Web sites for locating online communities are listed in the Additional Resources section at the end of this chapter (see also chap. 7, this volume). Ethical considerations (see chap. 16, this volume) suggest caution about intruding on the privacy of people who did not join a community to become a research participant. Researchers may do well to observe a community for a while to get a sense of whether its members would be receptive to an invitation to participate in research (Eysenbach & Till, 2001).

The more passive technique of waiting for participants to discover your Web-based questionnaire does not mean doing nothing. It is important to design your pages so that they can easily be indexed by major search engines such as Google. General good advice for optimizing your pages for search engines and specific guidelines for registering with

Google, Yahoo!, and MSN can be found at http://www.searchengines. com/. Web page optimization increases the chance that individuals who search for key phrases such as "anxiety research" will find your site because it was properly indexed by the major search engines. In addition to registering your Web site with the major search engines, you should also consider listing it with Krantz, who maintains an up-to-date list of online research studies at http://psych.hanover.edu/Research/exponnet.html.

Sometimes concerns are expressed about whether one can obtain "representative" samples from the Internet by any recruiting strategy. This concern is usually misguided. Aside from the fact that most Internet samples are at least as diverse as the frequently used college student samples (Gosling, Vazire, Srivastava, & John 2004), obtaining a representative sample is unnecessary for research that tests hypotheses about relations among variables, unless one has reason to believe that participants in the sample have unique characteristics that interact with the variables being studied (Leary, 2004). Sampling is more an issue for survey research (see chap. 12, this volume), when one wants to accurately describe the thoughts, feelings, or behavior of a particular population, in which case the sample needs to represent the population.

DESIGN THE FRONT PAGE

The front page (or home page) for your Web-based self-report serves two functions. The first function is to persuade potential participants to complete your measure. Persuasiveness is particularly important if you are using a passive recruitment strategy, but even if you are actively directing participants to your site, the front page should be welcoming rather than forbidding. Fraley (2004; pp. 285–289) offered good advice concerning effective design for a front page. Here I emphasize just two of Fraley's points. First, you must strike a balance between making the page look professional and scholarly yet flashy enough to be eye-catching. The result should appeal to potential participants' desire to learn more about themselves. A good example of such a site is the popular and successful http://www.outofservice.com. Also, your front page should be a base from which participants can branch to any of a number of online studies that you or your lab group is currently conducting. This way, as studies come to completion, participants will still be able to find your site.

The second important function of a front page is to obtain informed consent from participants. Requirements for obtaining informed consent differ across universities and agencies, so the details will depend on the regulations of the organization that is supporting your research. For example, for the self-report scales used at http://www.outofservice.com, the front page contains a link to the site's privacy policy, and the pages for different scales contain a link labeled "Read our consent form" that directs participants to a very short and simple explanation of risks and benefits. By completing the scales, participants are giving implied informed consent. Other organizations require researchers to channel participants through a more involved, legalistic Web page and have participants actively give informed consent by clicking a button before they can complete the scale. For examples, see Revelle's form at http://test. personality-project.org/survey/consentform.html and one of my forms at the supplementary Web site for this chapter.

WRITE FILES FOR PRESENTING ITEMS

Two kinds of files can be used to present Web pages to your participants. The first is a file that contains fixed content and structure (defined by HTML tags) such that it always looks the same. I refer to these files as *straight HTML files*. The second type of file is a *script file*. Script files are written in programming languages such as Perl, PHP, and JavaScript. Script files can receive data, do calculations, and generate HTML files based on those calculations. This chapter shows examples of script files written in Perl, as this is the scripting language used in Fraley's (2004) introduction to Internet research methods. Again, readers unfamiliar with HTML should read chapters 3 and 4 of Fraley's book, and readers unfamiliar with Perl should read chapters 5 and 6.

Your front page, which represents the entry point for participants completing any of your self-report scales, should be a straight HTML page with a fixed URL so that search engines can find and index your site properly. Typically, this page will not contain self-report scale items. Rather, it will contain short descriptions of different scales with links that take you to other Web pages that contain the informed consent forms for the scales or to the scale itself (when informed consent is handled by a link to a separate page). For example, http://www.personal.psu.edu/ ~j5j/IPIP/ is a front page that directs people to either a 300- or 120-item IPIP representation of the Revised NEO Personality Inventory (NEO PI-R; Costa & McCrae, 1992); this representation is referred to as "IPIP-NEO" hereinafter.

My IPIP Web site is currently designed to educate visitors about personality assessment rather than to test research hypotheses, so following either link from my front page takes people to a disclaimer page rather than an informed consent form. Nonetheless, this disclaimer page illustrates two features that one might use on an informed consent Web page. One is a pair of checkboxes that must be ticked to access the actual self-report scale. The first checkbox requests the participant to accept responsibility for the time invested in completing the inventory, regardless of the results. The second checkbox asks the participant to acknowledge an understanding of the limitations of the results. The precise wording adjacent to the checkbox for your own informed consent page would be whatever is required by your human subjects committee (see chap. 16, this volume). The checkbox is a useful tool to increase the probability that participants will actually read a consent form or any instructions or information you would like them to read.

The second feature I use on my disclaimer page that can also be used in an informed consent Web page is the HTML interactive form, which will send the status of the checkboxes to a script file named "shortipipneo1.cgi," which will either begin presenting the self-report scale items (if the boxes are checked) or halt and send a warning to the participants that they must give consent by checking the boxes to begin the self-report scale.

To see the disclaimer page for the 120-item IPIP measure, please visit the supplementary Web site for this chapter. You can see the warnings sent to participants by clicking the "Send" button without ticking the checkboxes. To see the underlying HTML code that generates the checkboxes and interactive form, use the "View Source" command in your browser.

Once your script file has verified that the person has given informed consent or has read the preliminary instructions, it will next generate HTML to present self-report items. Choices for presenting the items and possible uses include

- text boxes or text areas (for entering a code name or for responding to open-ended items),
- checkboxes (for *yes–no* items such as those on a mood or activity checklist),
- pull-down menus (for choosing one item from a list, such as ethnicity or country of origin), and
- radio buttons (for responding to items on a scale of, say, 1–5).

The HTML generated by shortipipneo1.cgi includes text boxes for recording a nickname and the person's age, radio buttons for indicating whether the person is male or female, and a pull-down menu for indicating one's country. The Perl code for generating this HTML can be examined at the supplementary Web site for this chapter.

This chapter focuses on the presentation of self-report items where responses are recorded on a 1–5 scale with radio buttons. The format I describe here presents a limited number of items (no more than 60) per Web page in an HTML table. Presenting more than 60 items at a time may overtax some systems, causing a program crash (Johnson, 2001).

Some Web designers have criticized the use of tables to control the layout of content on a Web page, arguing that this should be accomplished instead with Cascading Style Sheets (CSS). Nonetheless, the ease of using tables can outweigh the alleged advantages of CSS (Budd, 2004). This is especially true when the table is simple, as in the case of laying out items and response options in a self-report scale. Furthermore, whereas table critics are usually referring to borderless tables used to control layout, for self-report scales I use borders to visually connect each item to its radio button responses. With this visual connection, participants are less likely to accidentally use the radio buttons for the previous or following item. Items appear in the leftmost column of the table, and five radio buttons with the response anchors appear in the five adjacent columns. Placing the anchors immediately above all radio button rather than only at the beginning of the scale ensures that the response option meanings will remain visible as the participant scrolls down the page.

If you look at the HTML in shortipipneo1.cgi for generating items, on the supplementary Web site for this chapter, you will see that all five radio buttons for the first item have the same name, Q1, but that the value assigned to the buttons varies from 1 to 5 to indicate how accurately the item describes the participant. Reverse-scored items can be assigned decreasing values from 5 to 1, if desired. Alternatively, responses can be reversed later when scale scores are computed. One advantage to not reverse-scoring items at the level of collecting data is that it facilitates the counting of consecutive identical responses, which can be a sign of non-attentiveness to content (Johnson, 2005). This is explained further in the next section on error checking.

After the HTML for the 60th item, the table is closed with the table end tag, "</table>," and followed by HTML for a "Submit" button with a message that clicking the button will take the user to the next 60 items. The beginning of shortipipneo2.cgi, the script file to which the participant's responses are sent, contains the first of several error-checking routines that are described in the next section. The Perl code in shortipipneo2.cgi can be examined on the supplementary Web site for this chapter.

WRITE SCRIPT FILES FOR ERROR CHECKING

One elementary error researchers will want a script to catch is when a participant fails to provide required information. The IPIP-NEO produces a narrative report based on sex and age norms and may eventually use norms based on nationality, so responses to these demographic items are required. The line of Perl code in the shortipipneo2.cgi script that checks whether participants have identified themselves as male or female reads:

if ($Sex ne "Male" && $Sex ne "Female").

If the statement is true (i.e., the person has not identified as male or female), the program stops and prints a message indicating the need to provide that information on the previous page in order to continue. Similar checks are conducted for the age and nationality variables. One could also check to see that a participant has completed each item in the scale, although you might only advise participants about missed items rather than requiring them to complete all items, as human subjects committees sometimes want to give research participants the option of not responding to individual items.

An alternative to checking for missing responses in a subsequent Perl script is to use JavaScript within the original Perl script. When a participant clicks the "Submit" button, a JavaScript form validation routine checks for missing responses and presents a message either on the page or in a pop-up window informing the participant about the incomplete items. Fraley (2004) advised against using JavaScript because of incompatibilities across different browsers and because many people disable JavaScript to avoid annoying pop-up ads. Also, there are ways to get past JavaScript checking, so if you are serious about catching missing responses, you will have to use techniques such as the Perl illustration above in addition to using the JavaScript. Nonetheless, interested readers can search the Web for "JavaScript form validation" for more information on this topic.

If you have given participants unique identifiers to match data over multiple sessions or with other sources of information (e.g., from informants; see chap. 11, this volume), you will want to ensure that the identifiers (and passwords, if you are using them) are the ones that you have assigned participants. This can be accomplished with a script provided by Fraley (2004, chap. 12). If you do not need to actually track participants but you want to associate some kind of nickname or identifier with them, you can use a subroutine written by Hunt (1999) that generates a semi-random but pronounceable string of any length. My script, shortipipneo2.cgi, calls this subroutine to create a 23-letter identifier when a participant leaves the nickname box blank. This allows me to check for duplicate submissions (Johnson, 2005). The script also shortens user-generated nicknames longer than 23 characters to a 23-character identifier, and it pads nicknames shorter than 23 characters to a 23-character string. This maintains lines of uniform length in the file where data are saved.

Other kinds of error checking can take place only after all of the items in the self-report have been answered. In my own research, I have used these error-checking techniques after downloading completed protocols from all participants (Johnson, 2005), but one could easily use them as part of the standard scoring procedures, to which I turn next.

WRITE SCRIPT FILES FOR SCORING SCALE AND RECORDING THE RESULTS

Although Perl was created primarily for the manipulation and formatting of text, it does contain sufficient computational features to score a self-report scale, test for statistical anomalies, standardize scores, and graph the results. To accomplish these activities, the script files that present the scale items must also pass the responses entered by participants to subsequent script files until the final script file does the scoring and statistical analyses. Passing inputted data to subsequent scripts is achieved with the "Input Hidden" tag (Fraley, 2004).

After shortipipneo2.cgi receives the input from shortipipneo1.cgi, it generates the HTML for the final 60 items. When the participant responds to these items and clicks the "Submit" button, shortipipneo2.cgi passes, by way of "Input Hidden" tags, the demographic information and responses to all 120 items to shortipipneo3.cgi, which reads the information, stores responses to all 120 items in a data file for creating norms, scores the IPIP-NEO, standardizes the scores, and provides feedback to the participant. This script, which can be examined at the supplementary Web site for this chapter, uses essentially the same process described in chapter 5 of Fraley (2004) for saving data to a file, so that procedure is not described here. The remainder of this section describes the process of scoring and standardization. It also describes some final error-checking routines that could be implemented, if desired. The production of narrative feedback is discussed in the final section of this chapter. Space limitations preclude the reproduction of code from shortipipneo3.cgi in this chapter, so the reader may wish to print a copy of the code from the supplementary Web site to more easily follow the discussion that follows.

The 120-item IPIP-NEO is scored for 30 facet scales similar to the facets of the original NEO PI-R. These facet scores are then combined to yield five domain scores representing the major factors of the five-factor model (Costa & McCrae, 1992). Fraley (2004) showed how summing or averaging items into a single scale score is a very simple matter. One need only compute a total score variable with a Perl statement such as

$$\text{\$total} = \text{\$q1} + \text{\$q2} + \text{\$q3} + \text{\$q4} + \text{\$q5};$$

and the new variable, $total, will represent the sum of the five items. If any of the items (say, $q4) is a reversed-keyed item, and the values assigned to the radio buttons were not reversed, the proper value for such items is derived by subtracting them from the quantity (1 + the value of the highest possible response), which in this case would be (6-$q4).

For a multiscale inventory such as the IPIP-NEO, however, one can take advantage of looping in a script to score all of the scales with one short set of instructions when the items on all scales are separated by the same number of items. Five simple lines of code from shortipipneo3.cgi, labeled "# Score facet scales" can score all 30 facets because items on the same facet scale occur every 30 items on the inventory. For example, items 1, 31, 61, and 91 form the Friendliness facet scale. Values for these items are summed into the variable $ss[1] by incrementing the subscript of the array @Q (which contains all 120 item responses as variables Q[1] to Q[120]) by 30 three times (as $j increases from 0 to 3). The full loop is repeated 30 times, as $i increases from 1 to 30, generating 30 facet scores, $ss1 to $ss30, in the array, @ss. (For a review of scalar and array variables, see Fraley, 2004.)

For mnemonic convenience, shortipipneo3.cgi uses a similar looping procedure labeled "# Number each facet set from 1–6" to assign the 30 facet scores to new array variables, @NF, @EF, @OF, @AF, and @CF, referring to the Neuroticism, Extraversion, Openness, Agreeableness, and Conscientiousness facets, respectively. Finally, the facets scales are summed to yield the domain scales.

The standardization of the facet and domain scale scores is achieved in shortipipneo3.cgi under the label "# Standardize scores" by including several arrays of means and standard deviations for males and females of different age groups. The norm array chosen for an individual is based on an *if*-statement. For example, the *if*-statement that begins "if ($Sex eq "Male" && $Age < 21) {@norm =" defines the @norm array for male participants under 21 years of age. To make later computations simpler, a zero is placed as the first element in the array because Perl indices begin with 0 rather than 1, viz, $norm[0], $norm[1], $norm[2], and so on. The 10 elements in the first line of the array following the zero are means for the N, E, O, A, and C domain scores followed by the respective standard deviations for those scores. The next line contains means and standard deviations for the six facets of Neuroticism, and so forth. The variable "$id" is created to identify the comparison group (e.g., males under 21 years of age) for the narrative report.

The arrays for norms in shortipipneo3.cgi are based on an Internet sample of over 20,000 individuals (Johnson, 2005). One can create dynamic norms, however, by having a scoring script open a continuously updated data file, compute means and standard deviations based on the most recent data, and place these values into a @norm array. Fraley (2004, chap. 6) explained how this can be accomplished.

Next, shortipipneo3.cgi standardizes the raw scores into *t* scores by applying the usual formula, using the means and standard deviations in @norm. Again, the Perl script takes advantage of the systematic ordering of facets by using a loop. To present feedback to participants, the shortipipneo3.cgi script next estimates percentile scores from the standard *t* scores, on the basis of a cubic relation between the two distributions under normality. Discussion of the estimation is beyond the scope of this chapter. However, one could also compute actual percentiles on the basis of accumulated data—dynamically, if desired—and report results in those terms.

Before turning to the production of narrative feedback, I describe additional error-checking measures that one could implement, if deemed appropriate. In my own research (Johnson, 2005) I normally apply these algorithms with statistical software after data are downloaded to screen out duplicate and otherwise problematic protocols. For demonstration purposes, I describe below how these procedures can be accomplished in real time with Perl so that problematic cases are not even added to one's database.

Missing Values

Entire books (e.g., Little & Rubin, 2002) have been written about options for handling missing data, some of them quite complex. The optimal strategy may be to preserve the encoding of all missing values as zero (the default in HTML forms when no data are entered) for your data file, to be handled later when you conduct data analyses, while using a simple method for dealing with missing data (e.g., replacing missing values with the midpoint of the rating scale) when you score a scale for feedback. A suggested first step is to simply count how many items were skipped with an incremental counter. You can use your own judgment to determine how many items can be skipped before you refuse to score the scale. For the 300-item version of the IPIP-NEO, 75.6% of the 20,000+ protocols had fewer than 3 missing responses, and the frequency curve showed a sharp decrease in cases after 10 missing responses, suggesting a cutoff at that point (Johnson, 2005). For protocols with fewer than 11 missing responses, the missing values can be replaced with the rating scale midpoint, 3, and a warning about the replaced values can be included in the narrative report. (Remember that the script should save the entire protocol to the data file before the missing values are replaced.) The demonstration script file at the supplementary site for this chapter contains a routine to handle missing values following the label "# Count missing values."

Excessive Use of the Same Response Category

Especially for longer, multiscale measures, a respondent will occasionally ignore the content of items, either across the entire measure or for a set of items, and will simply choose the same response option repeatedly to get through the measure quickly. For my large Internet sample (Johnson, 2005), I plotted frequency curves for each response category, showing the number of respondents who used the same response category twice in a row, three times in a row, and so on, and identified breaks where the frequency dropped off sharply. This suggested maximal strings of 9, 9, 8, 11, and 9 consecutive uses of each of the five response categories. These values are identical or similar to the values observed by Costa and McCrae (in press) for their NEO-PI-R. Researchers with access to data from a large sample who completed their scale can use either method to set cutoffs; otherwise, they must set some arbitrary limits. Perl code that could be used to flag excessive use of the same response category appears in the demonstration script file at the supplementary Web site for this chapter under the label "# P array unreverse-scores the reverse-keyed items." (Note that because my Likert scales were set up to code reverse-scored items as 5–1 rather than 1–5, I had to create a new array of item responses, $P[1]$ to $P[120]$, with the reverse-scored items recoded to reflect the actual response category used by participants.)

Protocol Inconsistency

Without some minimal level of consistent responding to items, a protocol is invalid and uninterpretable. Item response theory (Reise, 1999) offers models of response consistency that could, in principle, be used to examine response consistency on Web-based self-report scales. However, these models may be too complex and computationally intensive to implement in Perl as a real-time screening mechanism. In Johnson (2005), I described two simpler methods for quantifying consistency of self-expression on self-report scales, one described by Jackson (1976) and one described by L. R. Goldberg (personal communication, June 20, 2000). Both methods require relatively long (more than 30 items) measures, and the Jackson method further requires that the measure be a multiscale inventory. Because the methods cannot be used for short scales and because the program code for implementing them in the IPIP-NEO is lengthy, these measures are described only briefly here. Further details can be found in Johnson (2005). I currently apply these consistency algorithms to my data after they are downloaded rather than online. However, for demonstration purposes, Perl code for computing these consistency indices for the 300-item IPIP measure can be found at the supplementary Web site for this chapter.

Jackson's (1976) method involves forming two half-scales from every scale. Scores on the half-scales are correlated with each other across all half-scale pairs and the result corrected with the Spearman–Brown formula. Although it takes many lines of program code to compute a correlation coefficient from the half-scale scores, sums of squares, and sums of products, the process is straightforward. The range of Jackson coefficients representing acceptable internal consistency is discussed in Johnson (2005) and can be used to issue warnings in the narrative report or to set flags in one's data file.

L. R. Goldberg's (personal communication, June 20, 2000) method requires that one first intercorrelate all item responses to one's measure from a relatively large existing data set. The 30 unique item pairs with the highest negative correlations are identified, preferably items that are not forward-scored and reverse-scored items from the same scale. These items therefore represent item pairs that people generally answer in opposite directions. To calculate an individual's tendency to be consistent with general response patterns, program code computes the correlation across item pairs for an individual. A highly negative correlation indicates consistency with general response patterns. Again, interpreting the magnitude of this correlation to identify insufficiently consistent responding is discussed in Johnson (2005).

Identifying Duplicate Protocols

Although not a problem for producing meaningful feedback to an individual, duplicate protocols are undesirable in the database one is building for research. Duplicate protocols can be prevented by insisting that participants obtain a user ID and password before participating, but this disallows for anonymous participation. Fraley (2004) suggested using an initial item in which participants indicate whether they have participated before. This method is used at http://www.outofservice.com. Unfortunately, it will not detect multiple submissions that occur with repeated clicks of the final "Submit" button or reloading previous pages and resubmitting (Johnson, 2005). My preference has been to detect duplicates after the fact by identifying protocols submitted with the same nickname within a short period of time, with a suspiciously large number of identical responses, or both. This method is described in detail in Johnson (2005). Nonetheless, researchers may decide to block duplicate protocols from entering the database with a Perl screening routine that opens the data file, compares all information in the current protocol to all existing cases, and decides whether the information is similar enough to reject the protocol as a duplicate.

Social Desirability and General Considerations About Invalid Responding

One long-standing question in assessment by self-reports is whether people respond to items honestly or whether they respond in a way that they consider to be socially desirable (Paulhus, 2002). Researchers eventually realized that there are individual differences in the motive to present oneself in a socially desirable fashion. Thus, social desirability scales were constructed as checks on protocol validity. Such scales are freely available at the IPIP Web site, http://ipip.ori.org, and can be incorporated into one's substantive research scales if social desirability is a concern. Evidence suggests, however, that social desirability distortion represents an insignificant threat to validity in most contexts (Johnson & Hogan, 2006) and that invalid responding of any sort on Internet measures is a trivial problem (Johnson, 2005).

The script's final action will be to produce feedback for the participant after the final script file scores a protocol; examines it for missing values, nonresponsive patterns, and possibly duplications with previous protocols; and stores the information in a data file. This chapter concludes with that topic.

WRITE PROGRAMMING CODE
FOR PROVIDING FEEDBACK

As far as one can tell, a primary motive that encourages people to complete Web-based psychological measure is a desire to learn more about themselves, to compare themselves with other people, or to think about themselves in new ways. Providing meaningful feedback to research participants is therefore one of the prime incentives researchers can offer potential participants. Possibly the most important factor in designing appropriate feedback is that, unlike many popular quizzes on the Internet, it is research based. A second factor that makes feedback useful is when it goes beyond the trivially obvious. Telling participants who strongly agree with many items such as "I love parties" and "I enjoy being around a lot of people" that they are sociable may be accurate, but it is hardly enlightening. More useful would be to explain some of the consequences of being highly sociable, as revealed by research.

The IPIP-NEO measures individual differences along the heavily researched five factors of personality. This allows the narrative feedback report to include relatively rich and definitive descriptions of the five factors and their correlates. Much of this information goes beyond lay knowledge of personality. If you are using a relatively new measure, your feedback will necessarily be less extensive and more suggestive than definitive. Regardless of the amount of available research findings for a scale, the professional integrity of the report depends on making research-based statements about the implication of scores (Johnson, 1996).

With this in mind, the IPIP-NEO narrative report begins with an explanation of personality terms such as *trait* and *percentiles*. It then stresses that high scores are not better than low scores and that low, average, or high levels for different traits can be helpful, neutral, or detrimental in different contexts. It also discusses measurement error and how to resolve questions of accuracy. This general explanation of measurement by self-report is followed by descriptions of what research has revealed about persons who score high or low on each trait. Only after a personality trait is explained thoroughly is the participant presented with his or her standing on the trait. Most of the narrative focuses on the nature of the traits rather than on personal description, such that the report reads almost like a personality textbook, instructing the participant about each trait. See a sample report for the IPIP-NEO or complete the measure for your own report at the supplementary Web site for this chapter.

The general introduction to personality assessment and the description of the five factors are the same in everyone's report. The portion of the report that is tailored to the individual participant's results is generated by Perl *if*-statements, as described in chapter 6 of Fraley (2004). For example, the variable $SE represents the *t*-standardized Extraversion domain score in shortipipneo.cgi, and the *if*-statements "if ($SE < $LO)," "if ($SE >= $LO && $SE <= $HI)," and "if ($SE > $HI)" determine whether a description of low, average, or high Extraversion, respectively, is printed. The variables $LO and $HI were set previously in the script at 45 and 55, such that scores within a half standard deviation of the mean are considered average. Hofstee (1994) explained why it is advisable to use only three categories along a continuum for feedback.

Currently the IPIP-NEO scripts at http://www.personal.psu.edu/~j5j/IPIP/ produce simple ASCII graphs of the participants' results, calibrated in percentile scores. A revision of the graphing routines is underway, using Fraley's (2004, chap. 11) method for creating pictorial bar graphs. The needed bar width is creating by taking the gray image file of pixel size 1 × 1 (Fraley, 2004, p. 182) and redefining its width as the standardized score multiplied by an experimentally determined constant. The demonstration script at the supplementary Web site for this chapter includes the Perl code for generating the pictorial bar graphs. The multiplier constant that produces bars of appropriate width was determined to be 4.1. Therefore, under the label "# Create graphs," one can see, for example, a width variable for Extraversion ($WEP) defined by multiplying the standardized percentile for Extraversion ($SEP) by this constant: ($WEP = 4.1 * $SEP). This width variable is then to determine the width of the image file, bargray.jpg, in the table row for Extraversion:

```
<img src='bargray.jpg' width=$WEP height='20'>
```

The production of graphs as well as verbal summaries can be an effective way to satisfy the desire of participants to know where they stand on traits relative to other people. They may already know from their own experience that they are relatively extraverted, but a bar graph scaled in percentiles can show them whether their level of extraversion puts them in the top 50% or top 10% of a reference group.

Summary and Conclusion

Web-based self-report scales are an efficient way to gather data on psychological traits and states for large numbers of research participants. The one disadvantage of this method, compared with paper-and-pencil measures, is lack of control over the assessment environment, but this is offset by the ability to detect nonresponsive behavior with routines in the scoring script. (Also, note that one can bring research participants into a computer laboratory to provide Web-based self-reports to enjoy the benefits of efficient data collection and scoring without losing control over the assessment setting.) With a Web-based self-report scale, scoring the scale, checking for errors, recording data, and providing customized feedback are completely automated and instantaneous processes. Purported drawbacks such as lack of diversity in samples and participants responding inappropriately turn out to be nonissues. Properly constructed front pages, registered with the major search engines, will draw large numbers of participants, and scripts can be written to detect nearly every kind of problematic protocol. The payoff for care and effort put into authoring good Web pages and CGI scripts for online self-reports will be large samples of high-quality data.

Additional Resources

RECRUITING FROM INTERNET COMMUNITIES

With the Internet in constant flux, it is impossible to list definitive, permanent sites for locating online communities, but the following three sites can be a good starting point:

- Google Groups: http://groups.google.com,
- Yahoo! Groups: http://groups.yahoo.com/, and
- LiveJournal Communities: http://www.livejournal.com/community/

To give a sense of what these resources will provide, a search on the keyword *shyness* at the time of writing this chapter turned up 140 groups on Google Groups, 116 groups on Yahoo!, and 113 communities on LiveJournal. Representative groups from each site with the number of members are, respectively, alt.support.shyness (848 members), shyness (1122 members), and social anxiety (948 members).

LEARNING SCRIPTING LANGUAGES

Professional programmers almost invariably write as if the reader already understands programming. Consequently, novices should always begin with Fraley's (2004) book. People usually find that once they have learned one programming language, it is much easier to learn additional languages. People also find that modifying existing scripts to their own uses facilitates their learning. Therefore, if after learning Perl you want to learn PHP, you might consider working with a generic PHP script for processing forms, made available by Göritz on her Web site, http://www.goeritz.net/brmic/. This page also contains links to additional sample scripts, information on other software and an introductory tutorial on the use of PHP, Apache, and MySQL.

RESOURCES ON THE SUPPLEMENTARY WEB SITE, HTTP://WWW.APA.ORG/BOOKS/RESOURCES/ GOSLING

- 120- and 300-item IPIP representations of the Revised NEO Personality Inventory (IPIP-NEO)
- Version of 120-item IPIP-NEO that incorporates built-in error-checking and pictorial bar graphs
- Plain text copy of the Perl code in shortipipneo2.cgi for error-checking
- Plain text copy of the Perl code in shortipipneo3.cgi for scoring, standardizing, saving data, and providing feedback
- Script for checking for response consistency with Jackson's method
- Script for checking for response consistency with Goldberg's method
- Sample IPIP report

References

Budd, A. (2004, May 12). An *objective look at table-based vs. CSS-based design*. Retrieved March 27, 2007, from http://www.andybudd.com/archives/2004/05/an_objective_look_at_table_based_vs_css_based_design/

Costa, P. T., Jr., & McCrae, R. R. (1992). *Revised NEO Personality Inventory (NEO PI-R) and NEO Five-Factor Inventory (NEO-FFI) professional manual*. Odessa, FL: Psychological Assessment Resources.

Costa, P. T., Jr., & McCrae, R. R. (in press). The revised NEO Personality Inventory (NEO PI-R). In S. R. Briggs, J. M. Cheek, & E. M. Donahue (Eds.), *Handbook of adult personality inventories*. New York: Kluwer Academic.

Eysenbach, G., & Till, J. E. (2001). Ethical issues in qualitative research on internet communities. *British Medical Journal, 323,* 1103–1105.

Fraley, R. C. (2004). *How to conduct behavioral research over the Internet: A beginner's guide to HTML and CGI/Perl*. New York: Guilford Press.

Goldberg, L. R., Johnson, J. A., Eber, H. W., Hogan, R., Ashton, M. C., Cloninger, C. R., & Gough, H. G. (2006). The International Personality Item Pool and the future of public-domain personality measures. *Journal of Research in Personality, 40,* 84–96.

Gosling, S. D., Vazire S., Srivastava S., & John O. P. (2004). Should we trust Web-based studies? A comparative analysis of six preconceptions about internet questionnaires. *American Psychologist, 59,* 93–104.

Hofstee, W. K. B. (1994). Who should own the definition of personality? *European Journal of Personality, 8,* 149–162.

Hunt, C. (1999). *random_password()* [Computer software]. Retrieved from http://www.extraconnections.co.uk/code/password

Jackson, D. N. (1976, November). *The appraisal of personal reliability.* Paper presented at the meetings of the Society of Multivariate Experimental Psychology, University Park, PA.

Johnson, J. A. (1996). *Computer narrative interpretations of individual profiles.* Unpublished manuscript, Pennsylvania State University, DuBois.

Johnson, J. A. (2001, May). *Screening massively large data sets for nonresponsiveness in Web-based personality inventories.* Invited paper presented at the Bielefeld-Groningen Personality Research Group, University of Groningen, the Netherlands.

Johnson, J. A. (2005). Ascertaining the validity of Web-based personality inventories. *Journal of Research in Personality, 39,* 103–129.

Johnson, J. A., & Hogan, R. (2006). A socioanalytic view of faking. In R. Griffith (Ed.), *A closer examination of applicant faking behavior* (pp. 209–231). Greenwich, CT: Information Age Publishing.

Leary, M. R. (2004). *Introduction to behavioral research methods* (4th ed.). New York: Pearson.

Little, R. J. A., & Rubin, D. B. (2002). *Statistical analysis with missing data* (2nd ed.). New York: Wiley.

Paulhus, D. L. (2002). Socially desirable responding: The evolution of a construct. In H. I. Braun, D. N. Jackson, & D. E. Wiley (Eds.), *The role of constructs in psychological and educational measurement* (pp. 49–69). Mahwah, NJ: Erlbaum.

Reise, S. P. (1999). Personality measurement issues viewed through the eyes of IRT. In S. E. Embretson & S. L. Hershberger (Eds.), *The new rules of measurement: What every psychologist and educator should know* (pp. 219–241). Mahwah, NJ: Erlbaum.

Simine Vazire

Online Collection of Informant Reports

nformant reports are ratings collected from participants' acquaintances, usually to provide an additional perspective or as a substitute for self-reports when participants are unable to rate themselves. In this chapter, I describe how the Internet can be used to collect informant reports. I provide recommendations at several stages of the process, including how to recruit informants, how to contact informants, how to administer the questionnaire, and how to deal with nonresponders, all the while paying special attention to increasing response rates and maintaining the validity of the data.

Much research can benefit from the added perspective provided by informants. In particular, researchers working with populations that are unable to give valid self-reports, such as young children, people with memory disorders, or people with personality disorders involving lack of self-awareness, may be especially interested in improving their methods for collecting informant reports. Furthermore, the validity of personality assessments is greatly improved by aggregating ratings across multiple informants (Block, 1961). Therefore, most researchers interested in improving the validity of their assessments will likely be interested in collecting data from multiple informants. Informant reports can be combined with self-reports to obtain a multimethod assessment and to validate self-reports. Thus, even researchers who have not used informant methods in the past may be interested in learning more about this method.

Although informant reports are not a new method, the new technology featured in this book and applied in this chapter should allow researchers to obtain informant reports with less effort, fewer costs, and more success than do traditional methods (Gosling et al., 2004). It is likely that many researchers are aware of the use of informant methods but have chosen not to use them because of the perceived costs and efforts entailed. This chapter will hopefully convince such researchers that the availability of the Internet now makes informant methods a much more feasible option. Traditional paper-and-pencil methods for collecting informant reports have proven to be impractical and yield only moderate response rates. In contrast, informant reports obtained through the Internet yield response rates from 76% to 95% (Carlson & Furr, in press; Vazire, 2006; Vazire & Gosling, 2004; Vazire & Mehl, 2008; Vazire, Rentfrow, & Gosling, 2007). Furthermore, many of these informants were people who could not easily have come to the laboratory to complete informant reports—many were family members or hometown friends of college student participants. The quality of the informant data was high, with excellent scale reliabilities and good agreement among the informants and between the participants and their informants.

Collecting informant reports on the Internet provides the following benefits:

- Response rates are high, even without incentives;
- obtaining reports from knowledgeable informants who live far away is easy;
- the Web interface is less burdensome than paper-and-pencil methods;
- the cost and time involved in postal mailings or bringing informants to a laboratory are eliminated; and
- the Web interface eliminates data entry and facilitates record keeping.

Implementation of the Method

Collecting informant reports online is relatively easy. Below I outline the steps involved, make recommendations for improving response rates and preserving the validity of the data, and address some common misconceptions about Web-based informant reports.

STEP 1. CREATING THE WEB PAGES

Before conducting the study, the researcher will have to create a Web site to point the informants to. The Web site will consist of three pages: a welcome page, a click-through consent page, and the actual questionnaire. The welcome page may not be necessary but is recommended when the consent form is long or contains technical information (as required by your institutional review board [IRB]). The purpose of the welcome page is to explain the purpose of the study and of the informants' ratings. This page should be kept as brief as possible but should convey the following important points (ideally in bullet-point format):

- explain the purpose of the study;
- explain the importance and value of the informants' ratings;
- assure informants of the confidentiality of their responses;
- ensure that the informants know who they are rating and their ID number;
- provide a contact person in case they have any questions;
- inform them that the next page they see will be the consent form, and once they click through that they will be taken to the questionnaire; and
- give them an estimate of how long it will take to complete the questionnaire (make sure this estimate is accurate; many researchers overestimate the time it takes to complete a questionnaire—people often don't deliberate as much as researchers expect!).

A sample welcome page is provided in Exhibit 11.1. The specifics of the consent page will depend on your IRB's specific requirements. See chapter 16 (this volume) for information and advice about IRB requirements. A sample consent page is provided in Exhibit 11.2.

Finally, the questionnaire itself will include whatever items you choose. For guidelines on creating the actual questionnaire in HTML and CGI scripts, see chapter 3 (this volume) or Fraley's (2004) book on designing Web questionnaires. See also chapter 10 (this volume) for pointers on self-report questionnaires, many of which can also be applied to informant report questionnaires.

One common concern among researchers is that the informants will be discouraged by a long questionnaire. In my experience, informants do not seem deterred by questionnaires as long as 100 items. Although no empirical research exists on this matter, I suspect that the length of time it takes to complete the questionnaire is more important than the physical length of the questionnaire (i.e., number of items). It is my guess that if the appropriate steps are taken (as outlined in the following subsections), informants will be willing to complete questionnaires that take up to half an hour to complete. In general, I have

---

. EXHIBIT 11.1

Sample Informant Welcome Web Page

Welcome to the Web Site for the University Personality Study

As part of our personality study, we would like to collect information about our participants' personality from people who know them well.

In the first stage of our study, the person who nominated you (hereinafter referred to as "X") completed some questionnaires about how they see their own personality. With X's permission, we are asking you to provide us with additional information about X's personality. We believe that friends and family have a unique perspective on personality, and we hope you can tell us more about what X is like. Thus, we would like to know your impression of X's personality.

The rating form that follows will allow us to obtain a more accurate measure of X's personality. Before we get to the questionnaire, however, we want to make sure you understand that your answers are completely confidential. **<u>YOUR RATINGS WILL NEVER BE SHARED WITH X</u>**. In fact, nobody outside of the research team will ever see your answers, and we will never use your name or any identifying information when presenting our results. Be sure to keep your ratings of X confidential, and do not make your ratings in the presence of X.

We also ask that you answer as **honestly** and **accurately** as possible. In order to further our scientific research on personality, we are relying on you to provide us with as accurate a description of X as possible. Your ratings will not affect X's relationship to Washington University in any way; they will only be used for scientific research purposes.

The next Web page will ask for your consent to participate. After you give your consent, you will be directed to the questionnaire, which should take no more than 15 or 20 minutes to complete.

Before you proceed, make sure you know:

▪ Who you are rating (i.e., who "X" is)
▪ Your participant ID number

If you have any questions or comments, please feel free to e-mail us

Thanks!
The University Personality Research Lab
<u>Proceed to Consent Form</u>

---

found researchers' fears regarding informant cooperation to be overly pessimistic.

To link each informant to the target he or she is rating, each target will have an ID number, and informants should be given ID numbers that clearly link them to the target's ID numbers. For example, I typically assign informants the same ID number that I have assigned to the target participant they are rating, followed by a letter. This allows me to uniquely identify each informant and link his or her ratings to the target participant he or she is describing. Make sure to include a field for

EXHIBIT 11.2

Sample Informant Consent Web Page

University Personality Study Consent Form

The Psychology Department and the University require that all persons who participate in psychological studies give their consent to do so. A description of the study appears below.

This study examines how people see themselves, and how they are seen by others who know them well. In order to compare self-ratings to others' ratings, we are collecting personality ratings from people who know our participants well. Eventually, we plan to compare these ratings to determine how well people know themselves, and how they are seen by others. We will also be comparing the self and others' ratings to measures of behavior to determine whether self-ratings or others' ratings can predict behavior better.

One of our participants (hereinafter referred to as "X") has nominated you as someone who knows them well. On this Web site you'll find a short (10–20 minutes) questionnaire that we would like you to fill out about X. Please keep in mind that you will not be describing yourself; you will be describing X. If you do not know who nominated you for this study (i.e., who X is), please e-mail us before completing this questionnaire.

X WILL NOT SEE YOUR ANSWERS. All of your data will be kept strictly confidential. Only our research team will see your answers. All written and verbal accounts of the results will be anonymous.

Before you proceed to the questionnaire, remember that agreeing to participate does not commit you to anything. You can always withdraw at any time, choose not to answer any of the questions, or deny us permission to use your data. Your decision as to whether or not to participate will not affect your present or future relationship with Washington University.

Please note that this study is being conducted purely as part of an academic research program and has no commercial connections.

If, after reading the above information, you decide to participate in our study, please click on the "I agree" button below to indicate that you understand all of the above and that you give your consent to use the data you provide. Clicking on this button does not constitute a waiver of any legal rights. Please feel free to ask the research coordinators any questions you may have before or after completing the questionnaire by e-mailing us at svazire@artsci.wustl.edu

Clicking below will take you to the questionnaire. Please complete the questionnaire only once. Please be sure you know who you are rating (who X is) before continuing. Thanks for your participation.

I Agree

---

informants to report their ID number on the questionnaire, preferably at the bottom, right above the "Submit" button. It is also important to include code that will not allow the informant to submit the questionnaire without a valid value in this field (for how to make certain fields required, see Fraley, 2004). Finally, because the confidentiality of the informant reports is paramount, make sure that the Web site where you store the informant data is secure and password protected (see chap. 15, this volume).

STEP 2. RECRUITING INFORMANTS

The easiest and most straightforward way to identify informants who can provide well-informed ratings of your participants is to ask the participants directly. After explaining the purpose of the informant ratings to the participants, simply ask the participants to nominate people who know them well. Some IRBs may require that participants get permission from their informants before supplying researchers with e-mail addresses. In this case, be sure to tell participants ahead of time to obtain permission before coming to the experiment. Another common problem is that participants will not remember their friends' e-mail addresses or will write down an incorrect or illegible e-mail address. One way to avoid these problems is to collect this information on a Web-based form. This will allow participants to open another browser window to access their own e-mail and find their informants' exact e-mail addresses, which they can then copy and paste into the Web-based form. Of course, this is only possible if the researcher can provide a computer with Internet access during the experiment or have the participant complete this form from home before or after the experiment.

It is usually recommended to obtain at least two informant ratings in order to benefit from aggregation of multiple ratings. In my experience, people are willing and able to provide at least three informants who know them well and who would be willing to provide ratings. The only information that is required from the participants about the informants is their first name and e-mail address. It is important to make it clear to participants that the informants' ratings will be kept confidential and the participants will not receive feedback about how their informants rated them.

Several measures can be taken at this step to improve the likelihood of compliance on the part of both participants and informants. First, when explaining the purpose of the informant reports to the participant, the researcher should make it clear that the opinions of both the participant and the informants are valuable. The goal is to convey to the participants and informants that they are assisting the experimenter in an important scientific endeavor. For example, if the ratings are personality ratings, the experimenter can explain that the participants themselves and their close friends and family members are in the best position to provide the most accurate and in-depth assessments of the participants' personality, and you, the experimenter, would like to tap into this expert resource.

Another measure I take to improve compliance is to ask the participants to tell the informants to expect an e-mail from the researchers and to encourage them to complete the rating. It is likely that when the informants receive this e-mail, they will ask the target participant about it. Thus, enlisting the participants' help in encouraging the informants to complete the ratings is important.

Should Informants Be Compensated?

For certain kinds of research, compensating participants can increase response and retention rates (see chap. 14, this volume). In informant studies, however, there are two reasons not to compensate informants. First, as I mentioned previously, it is relatively easy to obtain very high response rates without compensating informants. Second, compensating informants introduces an incentive for participants to cheat (i.e., fill out their own informant reports) and for informants to provide nonserious responses simply to get the reward. Thus, I recommend that researchers avoid compensating informants unless they are asking the informants to complete a difficult or time-consuming task. This has the added benefit of eliminating the monetary costs of collecting informant reports.

Potential Limitations of Self-Selected Informants

The methods described here for recruiting informants focus on obtaining informant reports from participant-nominated informants. However, it is also important to point out that there may be serious limitations to this method of recruiting informants. Specifically, research suggests that self-selected informants are more likely to rate the target positively compared with informants that are chosen by the researcher or selected at random from a group of well-acquainted peers (Oltmanns & Turkheimer, 2006). These alternative methods involve selecting informants from a preexisting group of well-acquainted peers, such as a group of students who live, work, or take classes together; basic training groups in the military; and so on. However, Oltmanns and Turkheimer's (2006) research also suggests that self-selected informants may have more insight than other potential informants into anxiety-related or "internalizing" personality traits.

STEP 3. CONTACTING THE INFORMANTS

Although this is the most tedious part of the process, it is crucial that the researcher contact each informant individually and that the e-mail comes (or at least appears to come) directly from the principal investigator. Researchers may be tempted to ask participants to e-mail the informants themselves or to send a mass e-mail, but feedback from other researchers to me suggests that these shortcuts are costly in terms of response rates. Furthermore, sending out hundreds of e-mails does not take as much time as one might think.

In these days of junk mail filters, getting through to the informants can seem like a daunting task. Several steps can be taken to make sure the informants will receive the e-mail and to increase the chances that they will open it, read it, and follow the instructions. The first step, as mentioned above, is to ask the participants to tell their informants to expect an e-mail from the researcher. In addition, the researcher can take the following steps:

- E-mail each informant individually (this will be necessary anyway);
- use a university e-mail address (rather than one from Yahoo!, Gmail, Hotmail, etc.), and preferably one belonging to a faculty member;
- use the target participant's name in the subject line;
- address the informant by his or her first name in the first line of the e-mail; and
- keep the e-mail short, professional, and informative.

A sample initial e-mail to informants is provided in Exhibit 11.3. Note that the pronouns refer to a male target and would have to be changed for a female target. The labels in brackets refer to characteristics of the informant or target participant.

In a study with 100 participants and 3 informants per participant, 300 individual e-mails would need to be sent. This may seem like a lot of work, but with the help of the "Mail Merge" function in Microsoft Word, this can be accomplished in just 1 to 2 hours (and can be done by a research assistant if you are willing to give him or her access to your e-mail account—you can always change the password after he or she has completed this task). In addition, there are free programs available (e.g., WorldMerge by ColoradoSoft) that allow you to automate this process entirely. Specifically, WorldMerge (http://www.coloradosoft.com/worldmrg) allows you to use a Microsoft Excel spreadsheet as input and send out hundreds of individualized e-mails from your own e-mail account with the push of a button.

A slightly more tedious approach, but one that does not require any special software, is to create a template in Microsoft Word, then use the "Mail Merge" function to link it to a Microsoft Excel spreadsheet with all the information about the informants and target participants. The Excel spreadsheet should be set up as follows. Each informant is on a separate row. There should be one column for each of the following variables: informant e-mail, informant first name, target participant first name, target participant last name, target participant ID number, and informant letter. Recall that the informants will be identified by the target participant's ID number followed by a letter, providing a unique ID for each informant that also links them back to the participant he or she is rating.

EXHIBIT 11.3

Sample Informant Initial Contact E-mail (to an Informant for a Male Target)

Hello <informant firstname>,

<Participant firstname & lastname> recently participated in a psychology experiment here at Wash U, and he gave us your name as someone who might be able to help with our research.

As part of the experiment, we'd like to learn about what <participant firstname> is like, and he gave us your name as someone who knows him well and could tell us what he's like.

We have already asked <participant firstname> to tell us about himself, but to get another perspective on what <participant firstname> is like, we'd like to ask you to fill out a brief (5–10 minutes) on-line questionnaire telling us how you see <participant firstname>. Below is a link to the Web site with the questionnaire.

Please use the participant number provided below to complete the questionnaire.

Your participant number: <participant number><informant letter (A/B/C)>
Web site: www.simine.com/rate

If you have any questions, don't hesitate to e-mail me back or call me at 314-555-1266.

Thanks for your help!

Simine Vazire, PhD
Department of Psychology
Washington University in St. Louis

Once the Microsoft Word template is created, you can simply click a button to fill in the variables from the Excel spreadsheet. You can then copy and paste the text of the document into an e-mail program, fill in the informant's e-mail and subject line, and send the e-mail. Then simply click through to the next informant in your Microsoft Word file and repeat the process. (Your wrist may be tired by the end of the process, but you will hardly have broken a sweat.)

STEP 4. FOLLOWING UP

Many of the informants may not complete the questionnaire right away, but researchers should not be discouraged by this. The trick now is to balance the benefits of reminding the informants frequently with the risk of annoying them and being relegated to the junk mailbox. To manage this, my strategy has been to send one reminder each week for 2 or 3 weeks after the initial e-mail. The follow-up e-mails are much shorter than the initial e-mail, with the initial e-mail copied at the bottom. A sample follow-up e-mail is included in Exhibit 11.4.

Of course, it is important to keep track of who has completed the questionnaire and not to send follow-up reminder e-mails to these people! This

---

EXHIBIT 11.4

Sample Informant Follow-Up E-mail

Hello <informant firstname>,

I'm writing to remind you about the online questionnaire about <participant firstname & lastname> that we wrote you about last week. I know you are probably quite busy, but if you could take a few minutes to complete the questionnaire, we would really appreciate it!

Here's the info you need:

Your participant number: <participant number><informant letter (A/B/C)>
Link to questionnaire: www.simine.com/rate

Below is the first e-mail I sent you with more information about our research.
Let me know if you have any questions!

Thanks,

Simine

---

can be done by sorting the informant data by ID number and matching these up to the ID numbers of the informants in the Excel spreadsheet that is used for contacting informants. Those who have already provided data should be moved to the bottom of the e-mail contact spreadsheet and skipped when sending follow-up e-mails. Also, if your research involves repeated contact with the target participants, you might consider asking them to remind and encourage their informants to complete the questionnaire. Finally, it is probably important to consider the timing of the e-mails to the informants. In my experience, the best time to contact informants is during the semester, avoiding midterm and final exam periods.

Despite all of these efforts, there will probably be some informants who will not complete the questionnaire. If this is because the e-mail could not be delivered (e.g., because of an error in the e-mail address or because the recipient's mailbox is full) or if a given target had fewer than 2 informants, I will e-mail the target participant and ask him or her to nominate another informant. My goal is usually to reach a 75% response rate and obtain at least 2 informants per participant.

Conclusion

Informant methods can be very useful for researchers, allowing them to address new research questions and to improve the validity of their existing assessment tools. Nevertheless, many researchers do not make use of this method, presumably because they feel that the costs are too high and that the response rates are too low. It is my hope that the availability of the Internet, and in particular the use of the methods described here, will help put informant reports within the reach of all researchers.

Additional Resources

The section on the supplementary site (http://www.apa.org/books/resources/gosling) for this chapter provides resources for collecting informant reports.

Fraley, R. C. (2004). *How to conduct behavioral research over the Internet: A beginner's guide to HTML and CGI/Perl.* New York: Guilford Press.

This how-to guide will teach you about conducting research on the Internet, including how to make a Web questionnaire, how to present stimuli, how to link responses from multiple Web pages, and so on.

Vazire, S. (2006). Informant reports: A cheap, fast, and easy method for personality assessment. *Journal of Research in Personality, 40,* 471–481.

This short article describes Vazire's method of collecting informant reports, along with results (e.g., response rate, self–other agreement, other–other agreement).

References

Block, J. (1961). *The Q-sort method in personality assessment and psychiatric research.* Springfield, IL: Charles C Thomas.

Carlson, E. N., & Furr, R. M. (2009). Evidence of differential meta-accuracy: People understand the different impressions they make. *Psychological Science, 20,* 1033–1039.

Fraley, R. C. (2004). *How to conduct behavioral research over the Internet: A beginner's guide to HTML and CGI/Perl.* New York: Guilford Press.

Gosling, S. D., Vazire, S., Srivastava, S., & John, O. P. (2004). Should we trust Web-based studies? A comparative analysis of six preconceptions about Internet questionnaires. *American Psychologist, 59,* 93–104.

Oltmanns, T. F., & Turkheimer, E. (2006). Perceptions of self and others regarding pathological personality traits. In R. F. Krueger & J. L. Tackett (Eds.), *Personality and psychopathology* (pp. 71–111). New York: Guilford Press.

Vazire, S. (2006). Informant reports: A cheap, fast, and easy method for personality assessment. *Journal of Research in Personality, 40,* 471–481.

Vazire, S., & Gosling, S. D. (2004). e-Perceptions: Personality impressions based on personal Websites. *Journal of Personality and Social Psychology, 87,* 123–132.

Vazire, S., & Mehl, M. R. (2008). Knowing me knowing you: The accuracy and unique predictive validity of self and other ratings of daily behavior. *Journal of Personality and Social Psychology, 95,* 1202–1216.

Vazire, S., Rentfrow, P. J., & Gosling, S. D. (2007). [Self-Reports, Informant Reports, Physical Appearance, and Judges' Ratings]. Unpublished raw data.

Tracy L. Tuten

Conducting Online Surveys 12

T his chapter provides researchers with the basic structure for designing and implementing online surveys. An *online survey* refers to a self-administered questionnaire that collects data from respondents using the Web as the mode of data collection (rather than, or in addition to, other modes such as mail and telephone). The sheer number of research companies offering online research services illustrates the strong affinity that exists today for Web-based survey research. The benefits are extensive, particularly in terms of resource use, including elimination of time and space boundaries, data entry, and postage and copying expenses. It is fast and inexpensive, especially when compared with the cost of mail and telephone surveys, with some costs often less than half those of traditional methods. There are challenges to face (and I discuss these shortly), but many problems can be solved through careful programming, data cleaning by hindsight, or both. And, most important, online surveys can solve some of the challenges of paper and telephone surveys. Some of the advantages of online surveys include the ability to use a variety of question formats (see chap. 3, this volume) and allow for multimedia (see chap. 4, this volume), pipe information from previous responses, use skip logic, and minimize some forms of response bias through randomization of questions and response options.

Online surveys are ideal for studies involving visual, video, and auditory cues (see chap. 4, this volume) because of the multimedia capabilities. Participants can see a range of visual and audio cues, such as print mock-ups and videos. Access to a dispersed group of participants with the capability to use and examine multimedia is an enormous advantage of online research to traditional research approaches. Online surveys also offer the potential to pipe information from previous answers into later questions, such that follow-up questions are possible and appear customized for the respondent.

Online surveys can also be programmed for randomization in the question order and presentation of response options, which minimizes the potential for response biases, a common concern in traditional survey approaches. It is very attractive for collecting data to use open-ended questions for two reasons. First, transcription is unnecessary, as the responses are received as electronic type. Second, responses to open-ended questions tend to be longer (and presumably more detailed) when collected through an online survey than when collected through paper survey. Furthermore, when requesting information on behavior and attitudes that may be subject to socially desirable responding, online surveys appear to make respondents feel more anonymous and therefore more honest in providing information to sensitive questions (Tuten, Urban, & Bosnjak, 2002). If the behavior and attitudes under investigation are related to the Internet, the online survey is a good option—the best way to understand responses to virtual environments is to conduct the study in that same realm.

Last, although still used only minimally, online surveys offer some insight into the behavior of respondents as they participate in the study. Metadata can reveal how long participants spent on various questions, how they moved through the survey, when and how often they visited the survey site, and more. Such information can be interesting in terms of the perceived involvement participants felt in the study, the potential difficulty they faced cognitively as they addressed various questions, and the information revealed relative to the quality of the data results (for a categorized list of online survey response behaviors, see Bosnjak & Tuten, 2001).

Clearly, there are many reasons to use online surveys as the data collection mode for a survey research project. Still, despite these benefits, survey data collection online will not be appropriate for every research situation. These limitations primarily have to do with potential sources of survey error (for greater detail, see Groves, 1989), including coverage error, sampling error, nonresponse error, and measurement error. The next section addresses these sources of error from the perspective of online surveys specifically.

Coverage Error

Coverage error refers to the potential that there are members of the target population who have no chance of being selected into the sample. For online research, the potential for coverage error is directly linked to Internet coverage. As Internet penetration increases, the risk of coverage error decreases. It is important to consider the target audience and whether online research is appropriate for that group. For instance, among audiences in Western countries, Internet penetration is quite high and coverage error is unlikely to be a primary concern. However, for some adult populations and in countries with less penetration, online research is still in its infancy. Coverage error need not be a major concern unless Internet access is not readily accessible for the population of interest.

Sampling Error

Sampling error, a measure of uncertainty in the sampling frame, is for many researchers a bigger issue than coverage. Although technically the issue of sampling error is no different for the Internet than for other research modes, the ability to minimize sampling error is perhaps more limited. This is because there is no list of Web users and no possibility to use an Internet version of random-digit dialing to select probability samples. IP addresses are unique to machines, not people; many people have numerous e-mail addresses, and having an e-mail address does not mean that the address is in use. Of course, the notion of sampling error only makes sense in the context of probability samples, and most online research is not based on probability samples.

Couper (2000) demonstrated that probability samples can be achieved through the use of list-based samples (arguably, the best approach), intercept surveys, prerecruited panels, and the use of the Web as one of multiple options in a mixed-mode data collection scheme. Of these, list-based samples and prerecruited panels of users have met with the most success but are still limited by access to lists of e-mail addresses. Knowledge Networks has successfully developed the ability to survey probability samples of the general population online. It has done so by recruiting via telephone and random digit dialing and placing WebTV in the households of those selected.

Unrestricted self-selected surveys, like entertainment polls (e.g., CNN polls) and volunteer opt-in panels (like that operated by Greenfield Online) are more common, in large part, because of sampling limitations. To respond to an unrestricted, self-selected online survey, an individual must find the survey through links or search engines, or if an individual knows the survey address, he or she can choose to go directly to the survey. The sample is self-selected by those individuals visiting the Web site. Thus, it is difficult to compare nonrespondents with respondents to ascertain key differences between the groups or to control the quality of the sample of respondents participating in the survey.

Nonresponse Error

Nonresponse error is the error that occurs when information is not obtained from all sample units. The risk associated with nonresponse is the potential for *nonresponse bias,* the bias that occurs when the nonrespondents differ substantially from the respondents. A reasonably high rate of response rates is perceived as the best line of defense against nonresponse bias.

In this regard, researchers considering the use of the Web as the mode of data collection should exercise caution. Response rates to online surveys are generally lower than response rates to comparable mail surveys. Panels report the highest response rates; for instance, Knowledge Networks has reported rates as high as 90% (Pineau & Slotwiner, 2009). However, it is important to remember that this is the response rate for specific surveys by those who have agreed to panel membership and met the necessary requirements. The reported rates do not disclose the percentage who refused panel membership, did not complete the initial registration, or were lost to attrition. Nonprobability surveys, such as unrestricted, self-selected surveys, often report high numbers of responses, but response rates cannot be calculated.

Why might nonresponse be an issue for online surveys? There are several reasons, including inaccessibility (despite "coverage," some may not be available during the survey administration period); technical artifacts, such as slow Internet connections, pop-up blockers, incompatible software, and disabled features; design effects, such as a confusing visual presentation within the survey; and noncompliance (i.e., the intention to refuse participation, to refuse to answer some questions, or to terminate the process prematurely). Guidelines for minimizing nonresponse are provided later in this chapter.

Measurement Error

Measurement error occurs when the participants' answers do not represent their true values on the measure. Groves (1989) pointed to several possible causes of measurement error, including respondent error (e.g., respondent's inability, lack of motivation, lack of opportunity) to answer correctly; error due to design features such as question wording, response options, question layout; and error due to the mode itself. Several design choices can influence the degree of measurement error in an online survey, including the survey format (scrollable vs. interactive), complexity of the design, number of items per screen, detail of instructions, the presence of progress indicators, framing, and presentation of response options.

Implementing Online Surveys as a Data Collection Mode

To know whether online surveys are a desirable mode, consider the following questions:

- Is it desirable that data collection be completed quickly and efficiently?
- Is your sample large and geographically dispersed?
- Does your study require the use of multimedia stimuli?
- Does the questionnaire involve skip patterns, the need to refer to an answer from a previous question, or both, to structure the wording for a subsequent one?
- Is there a desire to minimize response biases through randomization of question and response option order?
- Will you be including open-ended questions?
- Are you asking questions that might be perceived as sensitive in nature?
- Are you investigating phenomena that are Internet specific (such as the social influence of opinions posted on message boards)?
- Is Internet access ensured? and
- Is there a list of e-mail addresses for the target population, or is a convenience sample sufficient?

If you answered yes to these questions, online surveys are certainly a viable option for your study. Next, I look at specific steps in the process of implementing an online survey.

At this point, you have already completed several steps in the research process. You have determined the research problem, specified research questions and data needs, and established that a descriptive approach using a survey is appropriate. Moreover, you have assessed the possible coverage and sampling issues and concluded that an online survey is capable of reaching members of the target population (i.e., coverage is not a major limitation). You have chosen measures to be included (for more on self-report scales, see chap. 10, this volume) and created a questionnaire draft (see chap. 3, this volume) from which to work as the online survey is programmed.

What is left to be done? The remaining steps include establishing a timetable for survey execution, specifying the sample design (and the subsequent type of online survey to be used), writing the copy for sample contact, programming the survey and making any necessary design changes, pretesting the instrument, making revisions as necessary, launching the survey, initiating the various contacts with sample members and monitoring contact success rates, and monitoring the survey while in the field (see Table 12.1). The process can be organized using a timeline, which highlights the necessary steps and assigned timing for completion of each step. In particular, this is useful for planning contacts with sample members.

SPECIFY THE SAMPLE DESIGN AND ASSOCIATED TYPE OF ONLINE SURVEY

At this point, the researcher has already established the target population and assessed whether Internet coverage is an issue. If a nonprobability sample will be used, the choices include self-selected, unrestricted online surveys, nonprobability panels recruited though self-selection and promotion, and lists of e-mail addresses generated without the benefit of probability sampling. If using an unrestricted, self-selected online survey, the emphasis must be on recruitment to attract potential respondents to the survey site. Recruitment for unrestricted, self-selected online surveys may rely on banners located on frequently visited sites and on the use of incentives, but such techniques can be costly for a researcher. Click-through rates to banner ads for online surveys are often disappointing.

Lists, whether probability based or not, are desirable because of the researcher's ability to contact the sample members directly and to calculate response rates (and even assess nonresponse bias). Intercept surveys can also provide a probability sample. In the case of an intercept survey, a pop-up screen is used to invite prospects to participate in the survey and enable click-throughs to link to the survey site. Panels can also be based on probability samples, either of specific target populations or of the full population, as in the case of Knowledge Networks. If the researcher prefers to use a list, one must assess the availability of a

TABLE 12.1

Steps in the Development of an Online Survey

Stage	Tasks to complete	Timing considerations
Sample design	Establish need for probability or nonprobability sample Choose frame Assess availability of list and arrange for list rental if necessary	
Write copy for contacts	Prenotification Invitation Reminders	
Develop the online instrument	Consider design issues Program survey Revise as necessary	Depends on use of online survey software versus programming in HTML and survey complexity
Pretest the survey and revise as necessary	Once survey is in production, test every component rigorously Make revisions as needed and retest Pretest with a small selection of sample members prior to launch Deploy survey	Depends on pretesting specifications (completed in lab or in field), number of revisions required
Launch the survey; initiate contacts or recruit if necessary	If a list is available, deploy waves of contact per schedule If an unrestricted, self-selected approach is used, begin recruitment	Monitor survey responses to assess possible problems and timing of subsequent waves
Monitor survey while in the field Close the survey		Ongoing Establish date (which may depend on response rates)

list of e-mail addresses and arrange for list rental if the list is controlled by a list broker. Survey Sampling International is a good example of a source of e-mail addresses for online survey researchers (http://www.surveysampling.com).

WRITE THE CONTACT COPY

If you will be recruiting respondents through direct communication, the contact copy is an important consideration for online surveys just as it is for surveys conducted using other modes. It serves several functions including obtaining the respondent's cooperation, establishing informed consent (see chap. 16, this volume, for more on ethics when conducting online research), and enhancing response rates. Such contacts could be accomplished using direct mail, telephone contact, or e-mail, assuming the researcher has access to contact information, but for most online surveys, the mode of contact is e-mail. For this reason, the remaining information in this section assumes e-mail as the contact mode. You will need copy for the following points of contact:

- prenotification,
- invitation,
- reminder to nonresponders,
- reminder to partial completes,
- follow-up reminder to nonresponders, and
- follow-up reminder to partial completes.

The prenotification is quite simply that—it is a contact that serves to introduce the study and establish a relationship with the sample members. Dillman (2000) advocated the use of prenotification as a tool for improving response rates and studies examining their use support their efficacy.

The invitation for an online survey serves the same role as a cover letter in a mail survey. When preparing the invitation, consider the following guidelines:

- address the invitation to the prospective respondent;
- identify the organization(s) conducting the study to establish credibility;
- state the purpose of the study and its importance;
- give assurances of confidentiality or anonymity;
- explain how the prospective respondent was chosen;
- emphasize the importance of the response to the study's success;
- if an incentive will be offered, explain the terms of the offer;
- explain the study's time frame, including details on the closing date of the survey and incentive awards;
- note the approximate time commitment required to complete the survey;
- acknowledge likely reasons for nonparticipation to defuse them;
- embed a link to the survey URL;
- provide contact information for respondents who wish to inquire about the study, the researcher's qualifications, Institutional Review Board (IRB) approval, or any other issues related to the study; and
- include an advance "thank you" statement that assumes participant cooperation.

Reminders are used to generate additional responses and to maximize the completeness of responses received. They are sent to nonresponders and to responders who left the survey prior to completing it.

Consequently, separate copy is needed for these two groups. It is common to use two waves of reminders, so a slightly modified follow-up reminder should also be scripted. The second reminder should include the survey close date to initiate a sense of urgency (and scarcity) to the request. Reminders do not affect data quality negatively (Tuten, 2008), so it is clearly in a researcher's best interest to use them. Still, some researchers, in their quest for higher response rates, remind potential participants numerous times. There is certainly a diminishing marginal return to the continued use of reminders, and more important, there is a cost to all researchers when members of a target population are irritated by the overuse of reminders.

When writing copy for all of these contacts, carefully consider the subject line. Also, the "reply-to" address should be consistent with the server used to send invitations. Many Internet service providers establish spam filter parameters based on consistency between the "reply-to" address and the server initiating the e-mail contact. Likewise, the content of subject lines can affect deliverability as well. Monitor undeliverable e-mails to assess whether revisions in the subject line are warranted, to clean the list of sample members prior to future contact, and to use in reporting response rate specifications. In addition, monitor the e-mail address offered as a contact device to identify problems or concerns from those contacted.

DEVELOP THE ONLINE INSTRUMENT

Online instruments are programmed following a guide that covers the text for respondents including instructions, question wording, and response options. The guide documents provide programming instructions as well include the number of questions per screen and the grouping of questions, question format (e.g., matrix, drop-down box, open text box with number of allowed characters), skip patterns, and piping guidelines. Creating the draft in this way can result in fewer revisions once programming has begun because it ensures that the researcher has considered the design issues that may increase measurement error, item nonresponse, and survey abandonment. When developing the online instrument, follow these guidelines (further instructions for questionnaire design are presented in chap. 3, this volume; Dillman, 2000; and Tourangeau, Couper, & Conrad, 2007):

- use interactive surveys with multiple screens whenever the survey contains several items and it is important to measure response behaviors;
- use scrolling surveys or group several items per screen in interactive surveys when contextual information is provided in a series of questions;

- use scrolling surveys when the modem capacity is thought to be minimal;
- keep designs simple;
- use few items per screen (seek to minimize scrolling in interactive surveys);
- provide simple instructions;
- provide progress indicators, preferably indicators that reveal the estimated time left to complete (as opposed to percentage of survey left to complete);
- consider the effects of graphical framing on the respondent's perception of what is being asked;
- consider the effects of spacing and color on the perception of response options;
- force answers only when necessary (e.g., when an answer is required to pipe relevant information into a subsequent question);
- allow participants to leave the survey and reenter later;
- enable respondents to report any problems; and
- keep the instrument brief (completion time of 10 minutes or less if at all possible).

Surveys may be programmed using HTML code for those who are, or have access to, competent programmers (for a review of guidelines for programming online surveys in HTML and issues related to client-side programming vs. server-side programming, see Birnbaum, 2004). Increasingly, however, researchers are using the many off-the-shelf online survey software programs available. Many of the companies offering these programs also offer other services including the contact management (management of the list of sample e-mail addresses including deployment of prenotifications, invitations, and reminders and monitoring of deliverability rates and issues), survey hosting (so the researcher requires no server capabilities in-house), data management (housing, cleaning, and preparation of data), and data analysis. As one might expect, the more sophisticated the software and services, the more expensive the fees associated with the provider. A list of providers, ranging from inexpensive to expensive, is provided in the Additional Resources. When considering the trade-off between capabilities and cost, consider the following questions:

- Is it important for your sponsor or for credibility with sample members (or both) that your survey URL reflect the name of your affiliated organization? Some providers, e.g., SurveyMonkey.com, do not enable a customized survey URL.
- What services do you really need? Do you require contact management services, survey hosting, and data management? What can be managed in-house? Most providers enable researchers to choose a la carte from their offerings.

- What design capabilities exist with the survey software? In particular, assess whether skip logic, piping, question order randomization, response choice randomization, display of multimedia stimuli and randomized sequences of stimuli, and specific question formats (e.g., differential scales are not commonly available) are possible.
- How skilled must one be to program surveys using the software? Some programs, such as Confirm it require a higher degree of programming knowledge than others, like Inquisite.

PRETEST THE INSTRUMENT, MAKING REVISIONS AS NECESSARY

Once the survey is in production, pretesting can begin. The two stages of pretesting are (a) in-house testing of all design features and (b) pretesting with a small group of people who are reflective of the sample population. When testing in-house, use the following list of questions to ensure all aspects of the instrument are functional:

- Does the survey function correctly on both Macs and PCs?
- Does the survey design appear on the screen correctly regardless of the Web browser used (e.g., Firefox, Safari, Explorer)?
- How does the survey experience vary at different connection speeds?
- Is the survey logic in place? Do all skip patterns and piping function correctly?
- Do all limits on response choices function correctly? For instance, if the instructions indicate that participants should select only one response, is this limit enforced by the program and an opportunity to correct errors provided to participants?
- Can a sufficient amount of text be entered into text boxes?

Once all necessary changes are made, pretesting with a small subsample can begin. This may be conducted in a usability lab or in the field and may include cognitive interviewing to ensure that the measures, question formats, design decisions, and so on, are appropriate. Again, this stage should be used to identify methods of improving the survey, and any revisions identified should be made prior to launch.

LAUNCH SURVEY, INITIATE CONTACT, OR RECRUIT PARTICIPATION

When establishing the date for the survey launch, remember to consider whether there are events that may affect response to the survey. For instance, does the launch date coincide with a holiday or other event that may affect the timing or likelihood of response? Prior to launch, deploy the prenotification to the sample members (again assuming direct contact is possible). Once the survey is deployed, the invitation is delivered. Invitations may be staggered if the sample size is sufficiently large to warrant concern for managing servers. In fact, this may be a requirement for some service providers to ensure that servers are not overwhelmed by responses in a short period of time. Anecdotal evidence suggests that invitations be delivered mid-week to enhance the likelihood that recipients will read the invitation and respond.

Responses should be monitored, and when responses have slowed sufficiently, the first reminder should be deployed. Remember that there will be a reminder for those who did not respond at all and a different reminder for those who started the survey but did not yet complete it. Again, monitor responses, and when they have again slowed, initiate the second reminder. Subsequent reminders may be used, but use caution and courtesy in deciding whether to continue to contact nonresponsive sample members.

If a list or established panel is not being used, recruitment to the survey site will be necessary. Recruitment can be managed just as it would for any Web site. Prospective participants can find the survey site via search engines and links posted on other, ideally heavily trafficked Web sites. Advertising, including banner ads on relevant Web sites, high traffic Web sites, or both, can be used. Organizations can be asked to promote participation to their memberships. Publicity, when possible, can be an effective tool to generate hits to the survey site (e.g., a survey of attitudes toward public use of mobile phones, along with the survey URL, might be mentioned in a news story). Last, snowball recruitment can be used by asking those who find the site to share the link with others. Note that if advertising will be used, it must be designed and produced and included in any IRB approvals (for more on IRB issues for online research, see chap. 16, this volume).

MONITOR THE SURVEY WHILE IN THE FIELD

As the survey is in the field, monitor response patterns and e-mails to the e-mail address provided in the contacts to sample members and in the survey itself (e.g., many researchers include contact information on a privacy page that explains how data will be used and any safeguards taken to protect the data). Participants may initiate contact to point out problems in flow, logic, and data capture. Be on the lookout for server problems (e.g., if the server goes down during the fielding, an unplanned contact to recapture participants who were unable to access the site may be required), response numbers or rates, and the timing of reminders given the rate and patterns of response.

CLOSE THE SURVEY

On the established date or at the point at which no additional responses are being generated, the survey site can be closed. At that point, the study can shift from data collection to data management.

Conclusion

The purpose of this chapter was to inform readers of the advantages and limitations associated with collected data using online surveys and to provide guidelines for implementing online surveys. I emphasized the situations for which an online survey is an appropriate data collection mode and explained the sources of survey error against which online surveyors must guard. Furthermore, I explained the specific steps of the research process associated with executing an online survey. To learn more about online survey methods and survey solutions, additional resources are provided below.

Additional Resources

Some of the numerous online survey software programs and hosting services, ranging from little or no expense to substantial investment:

- Harris Interactive: http://www.harrisinteractive.com
- Greenfield Online: http://www.greenfield.com
- Confirmit: http://www.confirmit.com
- Inquisite: http://www.inquisite.com
- SurveyMonkey.com: http://www.surveymonkey.com
- Web Surveyor: http://www.vovici.com
- Zoomerang: http://www.zoomerang.com
- Perseus Survey Solutions: http://www.vovici.com

References

Birnbaum, M. (2004). Human research and data collection. *Annual Review of Psychology, 55,* 803–832.

Bosnjak, M., & Tuten, T. (2001). Classifying response behaviors in Web-based surveys. *Journal of Computer-Mediated Communication, 6*(3). Retrieved July 24, 2009, from http://jcmc.indiana.edu/vol6/issue3/boznjak.html

Couper, M. P. (2000). Online surveys a review of issues and approaches. *Public Opinion Quarterly, 64*, 464–481.

Dillman, D. (2000). *Mail and Internet surveys: The tailored design method* (2nd ed.). New York: Wiley.

Groves, R. (1989). *Survey errors and survey costs.* New York: Wiley.

Pineau, Vicki and Slotswiner, Daniel. *Probability Samples vs. Volunteer Respondents in Internet Research: Defining Potential Effects on Data and Decision-Making in Marketing Applications,* Knowledge Networks White Paper, Accessed July 24, 2009, from http://www.knowledge networks.com/insights/docs/Volunteer%20white%20paper%2011-19-03.pdf

Tourangeau, R., Couper, M., & Conrad, F. (2007). Color, labels, and interpretive heuristics for response scales. *Public Opinion Quarterly, 71,* 91–112.

Tuten, T. L. (2008). The effect of reminders on data quality in online surveys. In L. O. Petrieff. and R. V. Miller (Eds.), *Public opinion research focus* (pp. 1–8). Huntington, NY: NOVA Science Publishers.

Tuten, T., Urban, D., & Bosnjak, M. (2002). Internet surveys and data quality: A review. In B. Batinic, U. Reips, M. Bosnjak, & A. Werner (Eds.), *On-line social science* (pp. 7–14). Seattle, WA: Hogrefe & Huber.

Ulf-Dietrich Reips and John H. Krantz

Conducting True Experiments on the Web

13

W ith the success of the Internet, its wide proliferation, and the availability of Web application software for generating and conducting experiments on the Internet, there are now very good reasons to turn to the Web for experimental data collection. Above all, as Reips (2007) put it, the ultimate reason for using the Internet to conduct experiments is

> the *fundamental asymmetry of accessibility* (Reips, 2002b, 2006): What is programmed to be accessible from any Internet-connected computer in the world will surely also be accessible in a university laboratory, but what is programmed to work in a local computer lab may not necessarily be accessible anywhere else. A laboratory experiment cannot simply be turned into a Web experiment, because it may be programmed in a stand-alone programming language and lack Internet-based research methodology, but any Web experiment can also be used by connecting the laboratory computer to the Internet. Consequently, it is a good strategy to design a study Web based, if possible. (pp. 375–376)

This chapter covers methodological and practical information that helps researchers (a) identify how Internet-based experimenting can be a useful method for their research and (b) create and run their own Web experiments. We begin the chapter by reviewing differences and similarities between Web and lab experiments, because knowing these is of utmost importance in the decision to transport the experimental method to an online environment. Benefits, challenges, and solutions that lead to a checklist of standards applicable to Internet-based experiments are discussed. From there, we introduce several techniques that have been shown to be useful in implementing the standards, in allowing researchers to either use classic features of experimental design and procedure on the Web or to tackle issues specific to Web experiments. A section on the variety of technical approaches to conducting Web experiments follows, which includes a discussion of the use of media and other stimuli in Internet-based experiments. Near the conclusion, we provide recommendations for programs and editors and a list of important and useful Web sites that were created in support of experimental Web research. To maximize the chapter's utility, we focus on the application of specific tools and sites for all stages of experimental research, most of which are accessible through the iScience Server at http://iscience.eu:

- generating the experiment: WEXTOR (http://wextor.org; Reips & Neuhaus, 2002);
- recruiting participants: exponnet list (http://psych.hanover.edu/research/exponnet.html; by John Krantz), web experiment list (http://wexlist.net; Reips & Lengler, 2005), Web Experimental Psychology Lab (http://wexlab.eu; Reips, 2001), online panels (Göritz, 2007);
- and analyzing log files: Scientific LogAnalyzer (http://sclog.eu; Reips & Stieger, 2004).

We selected these tools because (a) they cover a range of needs for Web experimenters; (b) they are user friendly and many operate fully automatically; and (c) they are maintained by researchers with a track record in the field. We also selected them because we have extensively used them in our own Web-based research (e.g., Krantz, Ballard, & Scher, 1997; Krantz & Dalal, 2000; Reips, 2009; Reips & Funke, 2008; Reips, Morger, & Meier, 2001; Schwarz & Reips, 2001). Broader reviews of Web experiment strategies are provided by Reips (2000, 2002a, 2002b, 2007).

It is important to know differences and similarities between Web and lab experiments because currently most new Web experimenters have previous experience with laboratory research. Entering the realm of Internet-based research with this background, one is likely to be highly successful when dealing with the relationship between experimenter and participant, and with experimental methodology and with computers, but one may be prone to step into some traps typically related to the psychology of Internet use and Internet-mediated communication (see, e.g., Joinson, McKenna, Postmes, & Reips, 2007), changes to the difficulties and practicalities of what can be done via Internet or not, and underestimating technological variance of hardware and software on the Web (Reips, 2000, 2002a, 2002b; Schmidt, 2007). Many of the potential traps are related to the advantages and disadvantages of Internet-based experimenting that we discuss in response to the respective questions below, along with solutions to typical problems.

When designing a study, one must find an optimized balance between methodological advantages and disadvantages. We recommend including Internet-based experimenting within a multimethod perspective drawn from general concerns about overreliance on any single research method: validate findings with different methods in different settings. Design the study for the Web, and for comparison, run a subsample in a local laboratory—or even in two (for an example involving local laboratory groups in Vancouver, Canada, and Zurich, Switzerland, see Reips et al., 2001). Where differences between results from online and laboratory methods are found, there is often an obvious explanation in sampling, computer expertise, Internet expertise, or format (see chap. 3, this volume; the sections on disadvantages and techniques that follow; and Krantz & Dalal, 2000). Established recruitment practices from undergraduate student populations versus visitors of particular Web sites, for instance, may easily result in sampling differences that can alter response patterns (Reips, 2000).

Participation in Internet-based research is a form of computer-mediated communication, more specifically, Internet-mediated communication. Consequently, much of the research in this realm (e.g., Joinson et al., 2007) applies to the social psychology of the Internet-based experiment. After reading this chapter, you will know how to apply some of major issues related to this research.

Advantages of Internet-Based Experimenting

Conducting experiments via the Internet brings various benefits to the research (for summaries, see Birnbaum, 2004; Reips, 1995, 2000, 2002b). Main advantages are that (a) studies can be delivered to large numbers of participants quickly and with low effort; (b) when compared with laboratory research, Web experiments are more cost-effective in time, space, administration, and labor; and (c) one can recruit large heterogeneous or homogeneous samples, also of individuals with rare conditions of interest (Mangan & Reips, 2007; Schmidt, 1997).

Another advantage is that because of the anonymity of the setting, the Internet is particularly suited for studies on sensitive topics. For example, Mangan and Reips (2007) used the Web to reach and study people with sexsomnia. *Sexsomnia* is a medical condition in which one engages in sexual behavior during one's sleep. Difficult cases are highly distressing and have forensic implications. Sexsomnia may be quite common but often goes unreported because of embarrassment and shame. Thus, little is known about this condition's demographics and clinical features. Through the Web, however, it was possible to contact and survey more than five times as many individuals from this difficult-to-reach clinical population than could be reached in all previous studies from 20 years of research combined.

When discussing advantages of Internet-based experiments, sometimes the advantages of computerized assessment are mistakenly attributed to the new Internet method. However, many useful functionalities such as item branching, filtering, automatic checks of plausibility during data entry, and so on, were introduced to experimenting during the computer revolution in the early 1970s (see Drasgow & Chuah, 2006; Musch & Reips, 2000; Reips, 2002b). These functionalities are, of course, also available in Internet-based research and may combine well with Internet features. For example, depending on their IP address, participants can be routed to different Web pages (e.g., a researcher may want to create a different version for participants from the researcher's university). As with the research on sexsomnia mentioned previously, Internet-based methods facilitate research in areas that were previously difficult to reach or inaccessible (e.g., Bordia, 1996; Rodgers et al, 2001). Meta-analyses reveal that Internet-based methods are usually valid (e.g., Krantz & Dalal, 2000) and sometimes even generate higher data quality than laboratory studies (Birnbaum, 2001; Buchanan & Smith, 1999; Reips, 2000). Many other benefits of Internet-based methods are frequently listed (e.g., Birnbaum, 2004; Reips, 1995, 2000, 2002b, 2006; Rhodes, Bowie, & Hergenrather, 2003), among them

- *Participant-related advantages:*
    - the ease of access for participants (physically, psychologically, culturally) and
    - truly voluntary participation (unless participants are required to visit the Web site).
- *Methodological improvements compared with laboratory experiments:*
    - detectability of confounding with motivational aspects of study participation;
    - avoidance of time constraints; and
    - reduction of experimenter effects (even in automated computer-based experiments, there is often some kind of personal contact, not so in most Web experiments), in particular a reduction of demand characteristics.
- *Generalizability and external validity:*
    - better generalizability of findings (e.g., Horswill & Coster, 2001),
    - greater external validity through greater technical variance,

- generalizability of findings to more settings because of high external validity, and
- ease of cross-method comparison (compare with sample tested in the laboratory).
- *Visibility and public control:*
  - greater visibility of the research process (Web-based studies can be visited by others, and their links can be published in articles resulting from the research);
  - heightened public control of ethical standards.

Drawbacks of Internet-Based Experimenting

Potential disadvantages of Internet-based experimenting, on the one hand, reside in the traps mentioned initially that may catch a researcher who is habitually following certain laboratory procedures. On the other hand, disadvantages come with the Internet setting and the technologies involved. Experience has shown that frequently voiced concerns regarding Internet-based experiments, such as multiple submissions (thereby missing representativeness of Internet users) and dishonest or malicious behavior (false responses, "hacking"), are not as frequent and not as problematic as previously considered (Birnbaum, 2004; Birnbaum & Reips, 2005). Some of the real issues, such as interactions between psychological processes in Internet use and the widely varying technical context, tend to be overlooked, though, and are discussed below, along with solutions (Reips, 2002a, 2002b, 2007; Schmidt, 2007).

Computer anxiety, lack of experience with the Internet, or both, may lead to substantially different results for Internet-based studies than for studies administered on paper (Whitaker & McKinney, 2007). Buchanan and Reips (2001) were able to show that users logging onto an online version of the IPIP Big Five personality test (Buchanan, Johnson, & Goldberg, 2005) via a Macintosh computer scored significantly higher on Openness to Experience than did those using a PC. Also, they showed that people who had JavaScript turned off were more educated, on average. This means that the more complex the technology involved in programming a Web study, the more samples are likely to be biased demographically and psychologically.

Despite these findings, converging evidence shows that Internet-based research methods often result in qualitatively comparable results to traditional methods (e.g., Krantz & Dalal, 2000; Luce et al., 2007; cf. Buchanan et al., 2005), even in longitudinal studies (Hiskey & Troop,

2002). Some of the cases in which differences have been found may be explained by frequent configuration errors in Internet-based research (Reips, 2002a), which we present along the way as part of a step-by-step guide to conducting a Web experiment in the next section.

In the remainder of the chapter, we provide a brief user guide to conducting an Internet-based experiment. We do so by describing each step when using WEXTOR. This tool has been developed by pioneers of Internet-based experimenting, has many important techniques built in, and is constantly being expanded to integrate new ones. Because Internet-based experimenting can be markedly different from lab and field experimenting, Reips (2002b) proposed 16 standards or guidelines that may help researchers and reviewers of manuscripts that are based on Internet-mediated research. The first standard is to use such a Web-based software to create your experimental materials, because it implements procedures for Web experiments that guard against typical problems.

Using WEXTOR to Build a Web (or Lab) Experiment

We use WEXTOR to recreate the 2×2 factorial Experiment 1 about information leakage (McKenzie & Nelson, 2003; Reips, 2003). The Web experiment is on display at http://tinyurl.com/25ftae. WEXTOR was created by researchers to implement usability and advanced techniques, and to be low tech (static HyperText Markup Language [HTML] and JavaScript only). It is important to know that for your specific experiment, there may be alternatives: Check whether the advantages of dynamic HTML (PHP, Perl) or non-HTML scripting languages such as Java and plugins such as Authorware, Flash, or Shockwave outweigh their disadvantages. But it is important to realize that with more advanced methods, generally the participant population becomes more restricted.

Researchers log on to http://wextor.org (see Figure 13.1). They click on "Sign up" and provide a unique user name, password, and e-mail address, then receive an e-mail with an activation code that needs to be used just once after clicking on "Login." From then on, only login and password will be needed; free 6-month trial accounts are available. Once logged in, click on "Create/modify an experimental design." On the new screen, type a name for your new experiment (e.g., "cup"), then click "Create." Follow the steps below.

1. Define the types and number of factors in your design. Select "2" between-subjects factors, "0" within-subject, and "no" quasi-experimental factors for this experiment. Then click the

FIGURE 13.1

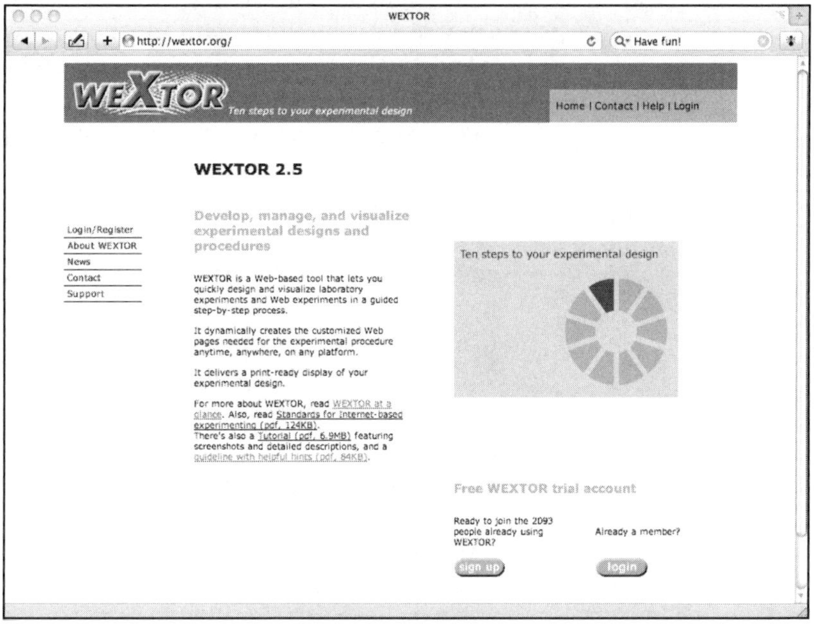

WEXTOR, the Web (and lab) experiment generator.

">" button to advance to the next step (do this after each step below).

2. Name the two factors "point of reference" and "order of answer options," or something similar. For each factor, type "2" where it says "Number of levels."

3. Name the levels "4-->2" and "0-->2" for "point of reference," "full first," and "empty first" for "order of answer options."

4. Name your experimental conditions. It is best to change nothing here; WEXTOR provides a naming proposal that avoids a frequently observed configuration error in Internet-based experimenting, namely, researchers' tendency to name their pages and folders too obviously (Configuration Error III; Reips, 2002a). Notice that the names proposed by WEXTOR follow Standard 11 (i.e., unobtrusive naming) recommended by Reips (2002b): They begin with a logical part that helps you identify the experimental condition (11, 12, 21, 22) and continue with three random characters that keep your participants from jumping pages or conditions. Obvious file naming like "a1.html, a2.html, a3.html, b1.html, b2.html" or "dissonance_study/control_cond/b3.html"

may reveal the experiment's topic, structure, or even the manipulation to participants.

5. Read about the standard pages included with each experiment. They serve some important functions, such as screening for a working Web browser, providing a form for informed consent, ensuring random distribution of participants to conditions, and so on. Add the number of additional Web pages you will need per condition.[1] It helps to get a piece of paper and briefly sketch the screens your participants will see to determine the number of pages needed. Think of one page per screen view and be advised that it is better to follow a "one-item–one-screen" philosophy rather than accumulating a lot of material on one page to measure dropout and time in a meaningful way (Reips, 2000, 2002b; chap. 3, this volume). Also, consider the warm-up technique and other types of drop-out control here that is explained at the end of this chapter. Do not worry if you pick the wrong number of pages; this can be changed later. For this experiment, choose "1."

6. Name your pages. It is best to keep the names WEXTOR proposes for the five standard pages and the suffixes, unless you would like to host the resulting experiment on your own server. You could also specify an automatic timing out for each of your pages. For this experiment, leave it at "0" and name Web page 1 "task" (.html).

Skip Steps 7 and 8; there are no within-subjects or quasi-experimental (natural) factors.

9. In Step 9a, view your experimental design either in list format (see Figure 13.2; great for methods sections in reports)[2] or in visual format (see Figure 13.3; great as a flowchart for the procedures section). Pick a style ("skin," implemented as a Cascading Style Sheet) for the layout of your pages, then open a preview in Step 9b by clicking "View." In 9b, you add content to your Web page by clicking on its name (do so now); editing of the standard pages is done later. First, choose "Plain text," then hit the "Add" button; a form appears. Leave the variable name as-is. Enter the following text to the second field:

---

[1] Note that WEXTOR does not provide an option for entering different numbers or names of pages for different conditions, because this would introduce a confound that has been shown to be of strong influence (see chap. 3, this volume). In exceptional cases, different page counts or names could be tolerated; the design is to be adapted by hand after download in Step 10.

[2] The list format also contains a code plan that you should review for variables and answer options once finished with all of Step 9.

FIGURE 13.2

Your experimental design

Your experiment consists of 2 factors:

Between-subjects factors

Factor 'point of reference'

- 4-->2
- 0-->2

Factor 'order of answer options'

- full first
- empty first

Experimental conditions

Experimental condition 1-1: 11cf69
- factor point of reference, level 4-->2
- factor order of answer options, level full first

Experimental condition 1-2: 12b006
- factor point of reference, level 4-->2
- factor order of answer options, level empty first

Experimental condition 2-1: 21a92d
- factor point of reference, level 0-->2
- factor order of answer options, level full first

Experimental condition 2-2: 222f7b
- factor point of reference, level 0-->2
- factor order of answer options, level empty first

Automatically generated list of design in Step 9a.

"Imagine a 4-ounce measuring cup in front of you that is completely filled with water up to the 4-ounce line. You then leave the room briefly and come back to find that the water is now at the 2-ounce line." The description field at the bottom is for your own notes; they will appear in the code plan in 9a. Confirm the entry, then choose "Radio buttons in vertical order" and hit "Add." Enter "What is the most natural way to describe the cup now?" in the second field and choose "2" answer options. On the subsequent screen, type "The cup is 1/2 full" and "The cup is 1/2 empty," confirm, and hit "view" in the upper right corner. A preview of the page will be shown in a pop-up

FIGURE 13.3

Visual display of your experimental design

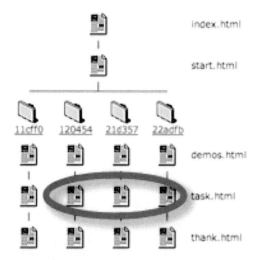

Automatically generated visual display of design and procedure in Step 9a. Pages resemble screens; folders resemble conditions.

window. Before you advance to 9c, note that two pieces have appeared under "Modify item." You could skip Step 9c by just advancing to Step 10, but you should note the default settings here could be changed to turn off client-side millisecond resolution response time measurement (standard server-side measurement of resolution in seconds is always provided), modify session ID length for enhanced control of multiple submissions, add *soft form validation* (i.e., the participant is warned if not all available form fields are filled in; this warning, however, is only shown once for each page to avoid provoking the participant to drop out), or add a high hurdle (Reips, 2000, 2002a).

10. Here, you download your complete experimental materials as a zip-compressed file. Decompress to your desktop; you now have a folder with the same name as your experiment (e.g., "cup") that contains five files, a folder for each condition and one for media files ("media" or "img"); see Figure 13.4. To test your experiment in its current state, open "index.html" in a

FIGURE 13.4

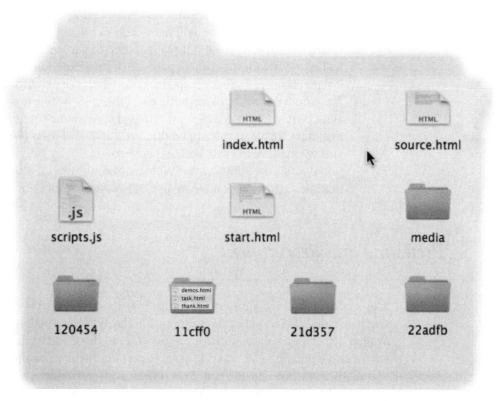

WEXTOR-generated experiment folder containing all subfolders, files, and scripts to run the experiment.

Web browser. Because WEXTOR creates static HTML pages, you can perform this testing and the next few steps even without being connected to the Internet.

11. One by one, load "task.html" from folders 12, 21, and 22 (marked in Figure 13.3) into an HTML editor (Dreamweaver, UltraEdit, BBEdit, NVU1.0, NotePad—but not MS Word; for an overview of HTML editors, see chap. 3, this volume) and change according to "Experimental conditions" in Figure 13.2. This means changing the instruction sentence in 21 and 22 to "... that is completely empty down to the 0-ounce line ... to find that there is now water up to the 2-ounce line." and changing the order of the answering options in 12 and 22. Make sure you exchange not only the text but also the radio buttons next to it, so "The cup

is 1/2 full" is consistently coded as "0" (or "1"). Feel free to also edit the demos.html and start.html pages—note the simple but powerful seriousness check technique (Reips, 2002a) on start.html (see Figure 13.5 and text below for an explanation). Answers to this question will allow you to later skip analyzing data from those who did not really intend to participate but still had a look at your experiment pages. Test your experiment while it remains on your desktop and make any necessary changes.

12. Zip-compress your folder with everything in it and log into WEXTOR. After login, click on "Upload an experiment, download data." Upload the zipped file; WEXTOR will then provide you with a link to your Web experiment. It is now fully functional and accessible on the Web around the world. The two next procedures to follow are pretesting and recruitment.

Pretesting and Recruitment

A procedure that is very important in Internet-based research is extensive pretesting. Pretest your experiment for clarity of instructions and availability on different platforms. Pretest in up to three waves: (a) Ask two or three people to take part in your experiment while you observe them; (b) send the link to your experiment to a few colleagues and friends only and ask them for feedback; and (c) once you begin recruiting real participants, soon check the data file for any problems.

FIGURE 13.5

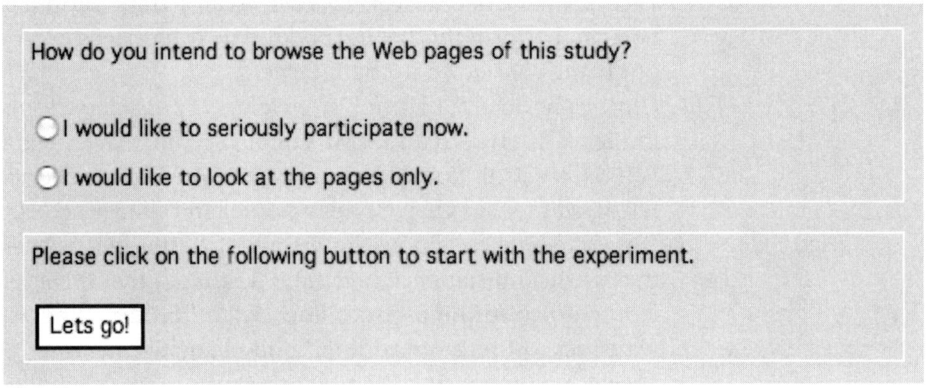

> How do you intend to browse the Web pages of this study?
>
> ○ I would like to seriously participate now.
> ○ I would like to look at the pages only.
>
> Please click on the following button to start with the experiment.
>
> [Lets go!]

Seriousness check technique.

PREPARATION FOR RECRUITMENT
OF YOUR PARTICIPANTS

Use the *multiple-site entry technique* (Reips, 2000, 2002a; for examples of successful implementations, see e.g., Horstmann, 2003; Janssen, Murre, & Meeter, 2007; Rodgers et al., 2003): For each source of recruitment (e.g., Web sites, forums, newsletters, e-mail lists, target groups) append a unique string of characters to the URL (e.g., ". . .index.html?source= studentlist"). Your data file will contain a column ("source") containing an entry of the referring source for each participant ("studentlist") for later comparison by subgroup (i.e., referring link). This technique provides you with indicators for effects of presence and impact of self-selection or contamination of samples. The technique may also be used to strategically segment a sample (Whitaker & McKinney, 2007).

The Web continues to be the primary route for recruitment (Musch & Reips, 2000; Reips & Lengler, 2005). Good places to announce one's Web experiments are the *exponnet site* and the *Web experiment list* (see Figure 13.6).

Once recruitment is completed, the data can be downloaded from WEXTOR, if WEXTOR was used for hosting of the Web experiment. Data are available in log file format (each access to an item—HTML page, image, and so on—is recorded in a separate line) that can be analyzed with Scientific LogAnalyzer and in semicolon-delimited data file format, suitable for import to spreadsheet software like Excel or SPSS. The data file contains all form data submitted by participants, the path taken through the experiment (e.g., showing whether the "Back" button was used), two different measures of response times per Web page, information about operating system and browser, and a quick view of dropout, among other information.

Techniques

WEXTOR automatically applies several techniques in the background that have been proposed for Internet-based experimenting (Reips, 2000, 2002a, 2002b) and promise to be useful to researchers and supportive to the method's quality. Some of these techniques have empirically been shown to work, for example, the seriousness check, compliance through early placement of demographic questions (Frick, Bächtiger, & Reips, 2001), the warm-up technique (Reips et al., 2001), double response time measurement (Reips, 2009), and the multiple-site entry technique (Hiskey & Troop, 2002).

FIGURE 13.6

The Web experiment list for recruitment and archiving of Web experiments. Currently, the searchable site contains more than 600 Web experiments, the sister site Web survey list contains more than 450 Web surveys.

TECHNIQUES OF EXPERIMENTAL DESIGN AND PROCEDURE

WEXTOR applied the following in your experiment:

- early versus late placement of demographic questions, resulting in lower dropout and better data quality (Frick et al., 2001), and
- meta tags to keep search engines away from all pages except the first page, so participants do not enter the study on one of the

FIGURE 13.7

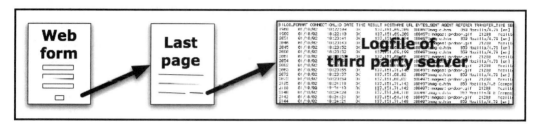

Illustration of Configuration Error II (Reips, 2002a). Study data may easily get transferred to third parties via the "referrer" information in the HTTP protocol, for example, via a link on a study's last page.

later pages and to prevent the study materials from being cached (good in case you make any changes to your pages):

> metaname="ROBOTS" content="NONE"><meta http-equiv= pragmacontent="no-cache">.

Reips (2002a) discussed several potential issues with Web experimental design and recommended the following techniques as solutions:

- Protection: Unprotected directories or the wrong choice of operating system and Web server combined with sloppy maintenance may compromise the confidentiality of participant data (Configuration Error I).
- Data transmission procedure: If data transmission occurs via the *GET method*,[3] then participants' form input collected from a Web page may be written to a third-party Web server log file (Configuration Error II). This will routinely happen if the link to the external Web server is two pages after the one with the critical form input, for example, on the last page of an Internet experiment (not unusual; see Figure 13.7). Solutions to the problem are links to other sites only where no data are forwarded to third-party sites and use of the POST method for data transmission.
- Technological variance: Reips (2002a) also discussed a general class of problems coming from the human tendency to ignore the substantial variance in technologies and appearances of Web pages in different browsers (Configuration Error IV). It should thus be standard procedure to pretest one's Web experiments on a variety of browsers on different operating systems.

---

[3] With the GET method, form data are appended to the URL of the next Web page that is called on by pressing the "Submit" button. Contrary to the past, when POST was inaccessible to the user, both POST and GET data format can be manipulated or switched client side in modern browsers by using add-ons like Tamper Data (https://addons.mozilla.org/de/firefox/addon/966).

■ Form design: Improper use of form elements (Configuration Error V), unfortunately, is a frequent problem in Internet-based research. For example, in drop-down menus, researchers trained in paper-based questionnaires design all-too-often forget to implement a preselected default option such as "please choose here." If such an option is missing, then each person who skips the question will be counted in for the first option in the list. A response option also considered important is something like "I don't want to answer" or "I don't know how to answer this"—such options will generally help in avoiding wrong or missing answers and therefore lead to better data quality.

For more details on meta tags, question order, and techniques of experimental design with regard to types and formats of *dependent variables* available in Internet research, see chapter 3, this volume. Chapter 3 also explains other issues of design and formatting in Internet-based research.

TECHNIQUES OF DROPOUT HANDLING (WARM-UP, HIGH HURDLE, REPORTING, DROPOUT CURVES)

Dropout (attrition) may be considered a problem in any study, even though it is important to keep in mind that dropout in Internet-mediated research can also be seen as an asset because the higher voluntariness on the Web creates enough variance to use dropout as a detection device or as a dependent variable. Dropout is particularly valuable in experimental research, in which different conditions are compared (Reips, 2000, 2002a, 2002b). Similar accounts can be made for other types of nonresponse (e.g., item nonresponse; Bosnjak, 2001). In this section, we describe several techniques that may be used, depending on whether the researcher intends to reduce, control, or use dropout or its influence.

Reips (2000) listed various background factors that are likely to influence participation and dropout, among them: (a) creating an attractive Web site, (b) emphasizing the site's trustworthiness, (c) providing a gratification, (d) offering feedback, (e) using a Web design that results in systematic shortening of loading times, (f) providing participants with information about their current position in the time structure of the experiment, and (g) using the high-entrance-barrier or high-hurdle technique.

The *high-hurdle technique* (Reips, 2000) is built on the assumption that participants vary interindividually in their motivation to participate in a Web experiment and that those with lower motivation are more likely to terminate their participation in case of difficulties. Instead of letting the lower motivated participants advance considerably into the study, a motivationally high hurdle is set at the beginning. Only moderately to highly motivated participants are expected to "jump" over it and continue with the study. Theoretically, the hurdle should move dropout from later pages to the hurdle and possibly increase data quality, because lower motivation to participate may also cause participants to respond carelessly.

The technique has been used repeatedly (e.g., Hänggi, 2004; Peden & Flashinski, 2004; Roman, Carvalho, & Piwek, 2006). When the technique is implemented through artificially increased loading times on the first study page, Göritz and Stieger (2008) recently showed that it indeed leads to a higher dropout at the hurdle, but later dropout and data quality appear to be independent of the hurdle.

A simple, but highly effective, technique is the *seriousness check*. Visitors to the first page of the experiment are greeted with the item depicted in Figure 13.5 that you saw on the start.html page in the experiment you created with WEXTOR. Musch, Bröder, and Klauer (2001), who first used this technique, placed the item at the end of their study. Experiments by Reips showed marked differences in dropout that can be predicted using the seriousness check item. If the nonserious answer is defined as an exclusion criterion, a large portion of dropout can be avoided. Experiments built with WEXTOR contain the seriousness check by default.

The *warm-up technique* avoids dropout during the actual experimental phase of a study by presenting the participants with tasks and materials before the experimental manipulation is introduced. Because most who drop out will do so on the first few pages of a Web study, their dropout cannot be attributed to the experimental manipulation if the warm-up technique is used (Reips, 2002a, 2002b). Dropout during the actual experiment was shown to be negligible (about 2%) in a Web experiments on list context effects by Reips et al. (2001).

Dropout is a valuable asset if used as a dependent variable, and it can be used as an indicator for *motivational confounding*, that is, a confounding of experimental manipulation with motivation, task difficulty, or both (Reips, 2000, 2002b). The issue is related to the voluntariness of and eagerness for participation in Internet-based experiments. In the typically remote, anonymous setting in which these experiments take place, there are fewer reasons that keep a participant from dropping out if he or she wishes to do so than in traditional laboratory experiments. There, the levels of a participant's motivation to engage in the experimental task might be confounded with levels of the independent variable, because those participants who are in the less motivating condition usually will not indicate so by leaving the situation.

By implementing a between-subjects design in the Web experiment, a difference in dropout can thus be used to detect such a motivational confounding. Furthermore, dropout curves can be used to detect differences in task difficulty or task attractiveness. Put as a general rule for experimenting: The less free participants are to leave the experimental situation, the less likely it is that motivation-related confounding variables will become salient (Reips, 2002b).

A Final Word on Technology

Internet-mediated research may take a variety of technological approaches, such as client side, server side, or a combination of both. Technologies may include HTML, XML, PHP, JavaScript, Java, Flash, and Shockwave, for example. Whichever technology is used on computers, there are several general problems with computer-based research, in particular on Windows (e.g., Myors, 1999; Krantz, 2000; Plant, Hammond, & Whitehouse, 2003; Plant & Turner, 2007). In addition, keep in mind that

- different versions of operating systems differ in vulnerability (Reips, 2002a);
- CRT and LCD monitors show considerable variation in intensity and color both between monitors and across the screen of a single monitor (Krantz, 2000, 2001);
- LCD monitors fare even worse in terms of timing accuracy (Plant & Turner, 2007); and
- computer mice, even of the same model, differ in timing (Plant et al., 2003).

Whenever such hardware and software issues affect an individual computer that is used as the sole device for an entire experiment, then the research is bound to be compromised. Thus, the Internet setting with participants all using their own computers is beneficial in this respect.

Many experiments demand the use of a stimulus that is manipulated in accord with the independent variable being studied. As such, the media capability of the Web is more important for experiments than many other types of research designs. Chapters discussed many of the issues related to the use of Web media for research purposes, and those issues are relevant here. Particularly, consider how media and the variation in access to or software for the media can cause confounds in an experiment. However, the low-tech principle discussed previously in this chapter also applies, perhaps even more strongly, to media. One of the primary ways connections vary is in media capability. As technology on the Web has advanced quickly, the ability of individual people on the Web to use that technology increases much more slowly, on average. The more important external validity is to the experiment, the simpler the media used should be.

We, for our part, favor the low-tech principle: Turn to solutions that do not need runtime environments or plugins, whenever possible. In plain text: Use HTML, XML, or PHP and maybe JavaScript. JavaScript seems to be sufficiently accurate and precise for millisecond reaction time measurement (Galesic et al., 2007; Reips, 2009; also see Reips et al., 2001), and it is built into every Web browser.

Additional Resources

META SITE

iScience Server (http://iscience.eu). This is a free portal to many of the services mentioned below. Hosted by Reips at the University of Deusto in Bilbao, Spain—after three years at the University of Zurich, Switzerland.

STUDY GENERATORS AND EDITORS

WEXTOR: http://wextor.org. This free Web application (with a small fee for comfort features), creates laboratory experiments and Web experiments and will work with between-subjects, within-subjects, quasi-experimental, and mixed designs. Many techniques important to Web research are built in. Its user friendly step-by-step process is also great for teaching students. WEXTOR stores and hosts the experimental designs, so you can work on and run your experiments using any type of browser from any computer.

FactorWiz: http://psych.fullerton.edu/mbirnbaum/programs/factorWiz.htm. This freeware generates HTML pages to conduct within-subject factorial experiments with random order of conditions.

idex: http://psych-iscience.uzh.ch/idex/index.html. This freeware is a Web service to create arcade-style game-based Web experiments. Data are immediately available for download.

VAS Generator: http://www.vasgenerator.net/. This free web service easily creates visual analogue scales for Web use.

Generic HTML Form Processor: http://www.goeritz.net/brmic/. This "Citeware" collects data from Web questionnaires using PHP.

Scientific LogAnalyzer: http://sclog.eu. This program is free for small log files in academic use and can analyze any type of log file. It also does dropout analysis.

EXAMPLES, RECRUITMENT, AND ARCHIVING

Decision Research Center: http://psych.fullerton.edu/mbirnbaum/ decisions/thanks.htm. Several student experiments that are research projects in judgment and decision making, supervised by Michael Birnbaum, one of the leading experts in Internet-based data collection. He writes on the Web page introducing the Decision Research Center: "As with any content you find on the Internet, after you link to a study, you should read the materials and decide for yourself if you want to participate."

Online Psychology Research UK: http://onlinepsychresearch.co.uk/. A Web site maintained by Kathryn Gardner at the University of Central Lancashire that is "designed to help researchers in the UK recruit UK participants (international researchers who wish to recruit UK participants are also welcome)." It limits participation to people for whom English is their native language.

Psychological Research on the Net ("Exponnet site"): http://psych. hanover.edu/research/exponnet.html. One of the most comprehensive Web sites listing online psychology related studies. The Web site is maintained by the second author at Hanover College.

Socialpsychology Network: http://www.socialpsychology.org/expts.htm. A Web site devoted to listing Internet-based data collection efforts specifically in the area of social psychology.

Web Experiment List: http://wexlist.net. Together with the "Exponnet list" mentioned above this is the largest list of Internet-based research studies on the Web. Researchers can fill in their own studies to be listed. Studies can be searched by category, language, status (active or not), and type (experiment or correlational study).

Web Experimental Psychology Lab: http://wexlab.eu. This was the first virtual laboratory with real data collection via experiments on the World Wide Web, founded in 1995 by the first author Then located at the University of Tübingen, Germany, it moved to the University of Zurich, Switzerland, in 1997, and since 2009 is located at the University of Deusto in Bilbao, Spain.

Birnbaum, M. H. (2004). Human research and data collection via the Internet. *Annual Review of Psychology, 55,* 803–832. This article is a comprehensive review of Internet-based research.

Reips, U.-D. (2002b). Standards for Internet-based experimenting. *Experimental Psychology, 49,* 243–256. The article reviews methods, techniques, and tools for Internet-based experimenting and proposes 16 standards.

Reips, U.-D. (2006). Web-based methods. In M. Eid & E. Diener (Eds.), *Handbook of multimethod measurement in psychology* (pp. 73–85). Washington, DC: American Psychological Association. The chapter discusses Web-based research in the context of multi-method measurement.

References

Birnbaum, M. H. (2001). A Web-based program of research on decision making. In U.-D. Reips & M. Bosnjak (Eds.), *Dimensions of Internet science* (pp. 23–55). Lengerich, Germany: Pabst Science.

Birnbaum, M. H. (2004). Human research and data collection via the Internet. *Annual Review of Psychology, 55,* 803–832.

Birnbaum, M. H., & Reips, U.-D. (2005). Behavioral research and data collection via the Internet. In R. W. Proctor & K.-P. L. Vu (Eds.), *The handbook of human factors in Web design* (pp. 471–492). Mahwah, NJ: Erlbaum.

Bordia, P. (1996). Studying verbal interaction on the Internet: The case of rumor transmission research. *Behavior Research Methods, Instruments, and Computers, 28,* 149–151.

Bosnjak, M. (2001). Participation in nonrestricted Web surveys: A typology and explanatory model for item nonresponse. In U.-D. Reips & M. Bosnjak (Eds.), *Dimensions of Internet science* (pp. 193–208). Lengerich, Germany: Pabst Science.

Buchanan, T., Johnson, J. A., & Goldberg, L. R. (2005). Implementing a five-factor personality inventory for use on the Internet. *European Journal of Psychological Assessment, 21,* 115–127.

Buchanan, T., & Reips, U.-D. (2001, October 10). Platform-dependent biases in online research: Do Mac users really think different? In K. J. Jonas, P. Breuer, B. Schauenburg, & M. Boos (Eds.), *Perspectives on Internet research: Concepts and methods.* Retrieved December 27, 2001, from http://www.psych.uni-goettingen.de/congress/gor-2001/contrib/buchanan-tom

Buchanan, T., & Smith, J. L. (1999). Using the Internet for psychological research: Personality testing on the World Wide Web. *British Journal of Psychology, 90,* 125–144.

Drasgow, F., & Chuah, S. C. (2006). Computer-based testing. In M. Eid & E. Diener (Eds.), *Handbook of multimethod measurement in psychology* (pp. 87–100). Washington, DC: American Psychological Association.

Frick, A., Bächtiger, M. T., & Reips, U.-D. (2001). Financial incentives, personal information, and dropout in online studies. In U.-D. Reips & M. Bosnjak (Eds.), *Dimensions of Internet science* (pp. 209–219). Lengerich, Germany: Pabst Science.

Galesic, M., Reips, U.-D., Kaczmirek, L., Czienskowski, U., Liske, N., & von Oertzen, T. (2007, September). *Response time measurements in the lab and on the Web: A comparison.* Paper presented at the annual meeting of the Swiss Society for Psychology, Zurich.

Göritz, A. S. (2007). Using online panels in psychological research. In A. N. Joinson, K. Y. A. McKenna, T. Postmes, & U.-D. Reips (Eds.),

The Oxford handbook of Internet psychology (pp. 473–485). Oxford, England: Oxford University Press.

Göritz, A. S., & Stieger, S. (2008). The high-hurdle technique put to the test: Failure to find evidence that increasing loading times enhances data quality in Web-based studies. *Behavior Research Methods, 40,* 322–327.

Hänggi, Y. (2004). Stress and emotion recognition: An Internet experiment using stress induction. *Swiss Journal of Psychology, 63,* 113–125.

Hiskey, S., & Troop, N. A. (2002). Online longitudinal survey research: Viability and participation. *Social Science Computer Review, 20,* 250–259.

Horstmann, G. (2003). What do facial expressions convey: Feeling states, behavioral intentions, or action requests? *Emotion, 3,* 150–166.

Horswill, M. S., & Coster, M. E. (2001). User-controlled photographic animations, photograph-based questions, and questionnaires: Three instruments for measuring drivers' risk-taking behavior on the Internet. *Behavior Research Methods, Instruments, and Computers, 33,* 46–58.

Janssen, S. M. J., Murre, J. M. J., & Meeter, M. (2007). Reminiscence bump in memory for public events. *European Journal of Cognitive Psychology, 20,* 738–764.

Joinson, A. N., McKenna, K. Y. A., Postmes, T., & Reips, U.-D. (Eds.). (2007). *The Oxford handbook of Internet psychology.* Oxford, England: Oxford University Press.

Krantz, J. H. (2000). Tell me, what did you see? The stimulus on computers. *Behavior Research Methods, Instruments, and Computers, 32,* 221–229.

Krantz, J. H. (2001). Stimulus delivery on the Web: What can be presented when calibration isn't possible. In U.-D. Reips & M. Bosnjak (Eds.), *Dimensions of Internet science* (pp. 113–130). Lengerich, Germany: Pabst Science.

Krantz, J. H., Ballard, J., & Scher, J. (1997). Comparing the results of laboratory and World Wide Web samples on the determinants of female attractiveness. *Behavior Research Methods, Instruments, and Computers, 29,* 264–269.

Krantz, J. H., & Dalal, R. (2000). Validity of Web-based psychological research. In M. H. Birnbaum (Ed.), *Psychological experiments on the Internet* (pp. 35–60). San Diego, CA: Academic Press.

Luce, K. H., Winzelberg, A. J., Das, S., Osborne, M. I., Bryson, S. W., & Taylor, C. B. (2007). Reliability of self-report: Paper versus online administration. *Computers in Human Behavior, 23,* 1384–1389.

Mangan, M., & Reips, U.-D. (2007). Sleep, sex, and the Web: Surveying the difficult-to-reach clinical population suffering from sexsomnia. *Behavior Research Methods, 39,* 233–236.

McKenzie, C. R. M., & Nelson, J. D. (2003). What a speaker's choice of frame reveals: Reference points, frame selection, and framing effects. *Psychonomic Bulletin & Review, 10,* 596–602.

Musch, J., Bröder, A., & Klauer, K. C. (2001). Improving survey research on the World Wide Web using the randomized response technique. In U.-D. Reips & M. Bosnjak (Eds.), *Dimensions of Internet science* (pp. 179–192). Lengerich, Germany: Pabst Science.

Musch, J., & Reips, U.-D. (2000). A brief history of Web experimenting. In M. H. Birnbaum (Ed.), *Psychological experiments on the Internet* (pp. 61–88). San Diego, CA: Academic Press.

Myors, B. (1999). Timing accuracy of PC programs under DOS and Windows. *Behavior Research Methods, Instruments, and Computers, 31,* 322–328.

Peden, B. F., & Flashinski, D. P. (2004). Virtual research ethics: A content analysis of surveys and experiments online. In E. Buchanan (Ed.), *Readings in virtual research ethics: Issues and controversies* (pp. 1–26). Hershey, PA: Information Science.

Plant, R. R., Hammond, N., & Whitehouse, T. (2003). How choice of mouse may affect response timing in psychological studies. *Behavior Research Methods, Instruments, and Computers, 35,* 276–284.

Plant, R. R., & Turner, G. (2007, November). *Precision psychological research in a world of commodity computers: New hardware, new problems?* Paper presented at the annual meeting of Society of Competitive Intelligence Professionals, Long Beach, CA.

Reips, U.-D. (1995). The *Web experiment method.* Retrieved January 6, 2009, from http://www.psychologie.uzh.ch/sowi/Ulf/Lab/WWWExpMethod.html

Reips, U.-D. (2000). The Web experiment method: Advantages, disadvantages, and solutions. In M. H. Birnbaum (Ed.), *Psychological experiments on the Internet* (pp. 89–114). San Diego, CA: Academic Press.

Reips, U.-D. (2001). The Web Experimental Psychology Lab: Five years of data collection on the Internet. *Behavior Research Methods, Instruments, and Computers, 33,* 201–211.

Reips, U.-D. (2002a). Internet-based psychological experimenting: Five *do*s and five *don't*s. *Social Science Computer Review, 20,* 241–249.

Reips, U.-D. (2002b). Standards for Internet-based experimenting. *Experimental Psychology, 49,* 243–256.

Reips, U.-D. (2003, September). *Seamless from concepts to results: Experimental Internet science.* Paper presented at the biannual conference on Subjective Probability, Utility and Decision Making (SPUDM), Zurich, Switzerland. Retrieved August 12, 2009, from http://www.psychologie.uzh.ch/sowi/reips/SPUDM_03/

Reips, U.-D. (2006). Web-based methods. In M. Eid & E. Diener (Eds.), *Handbook of multimethod measurement in psychology* (pp. 73–85). Washington, DC: American Psychological Association.

Reips, U.-D. (2007). The methodology of Internet-based experiments. In A. Joinson, K. McKenna, T. Postmes, & U.-D. Reips (Eds.), *The Oxford handbook of Internet psychology* (pp. 373–390). Oxford, England: Oxford University Press.

Reips, U.-D. (2009). *Reaction times in Internet-based versus laboratory research: Potential problems and a solution.* Manuscript submitted for publication.

Reips, U.-D., & Funke, F. (2008). Interval-level measurement with visual analogue scales in Internet-based research: VAS Generator. *Behavior Research Methods, 40,* 699–704.

Reips, U.-D., & Lengler, R. (2005). The Web Experiment List: A Web service for the recruitment of participants and archiving of Internet-based experiments. *Behavior Research Methods, 37,* 287–292.

Reips, U.-D., Morger, V., & Meier B. (2001). *"Fünfe gerade sein lassen": Listenkontexteffekte beim Kategorisieren* ["Letting five be equal": List context effects in categorization]. Unpublished manuscript, available at http://tinyurl.com/24wfmw

Reips, U.-D., & Neuhaus, C. (2002). WEXTOR: A Web-based tool for generating and visualizing experimental designs and procedures. *Behavior Research Methods, Instruments, and Computers, 34,* 234–240.

Reips, U.-D., & Stieger, S. (2004). Scientific LogAnalyzer: A Web-based tool for analyses of server log files in psychological research. *Behavior Research Methods, Instruments, and Computers, 36,* 304–311.

Rhodes S. D., Bowie D. A., & Hergenrather, K. C. (2003). Collecting behavioural data using the World Wide Web: Considerations for researchers. *Journal of Epidemiology and Community Health, 57,* 68–73.

Rodgers, J., Buchanan, T., Scholey, A. B., Heffernan, T. M., Ling, J., & Parrott, A. C. (2001). Differential effects of Ecstasy and cannabis on self-reports of memory ability: A Web-based study. *Human Psychopharmacology: Clinical and Experimental, 16,* 619–625.

Rodgers, J., Buchanan, T., Scholey, A. B., Heffernan, T. M., Ling, J., & Parrott, A. C. (2003). Patterns of drug use and the influence of gender on self reports of memory ability in ecstasy users: A Web-based study. *Journal of Psychopharmacology, 17,* 389–396.

Roman, N. T., Carvalho, A. M. B. R., & Piwek, P. (2006). *A Web-experiment on dialogue classification.* Paper presented at the Workshop on the Economics of Information Security (WEIS), Cambridge, England.

Schmidt, W. C. (1997). World Wide Web survey research: Benefits, potential problems, and solutions. *Behavior Research Methods, Instruments, and Computers, 29,* 274–279.

Schmidt, W. C. (2007). Technical considerations when implementing online research. In A. N. Joinson, K. Y. A. McKenna, T. Postmes, & U.-D. Reips (Eds.), *The Oxford handbook of Internet psychology* (pp. 461–472). Oxford, England: Oxford University Press.

Schwarz, S., & Reips, U.-D. (2001). CGI versus JavaScript: A Web experiment on the reversed hindsight bias. In U.-D. Reips & M. Bosnjak (Eds.), *Dimensions of Internet science* (pp. 75–90). Lengerich, Germany: Pabst Science.

Whitaker, B. G., & McKinney, J. L. (2007). Assessing the measurement invariance of latent job satisfaction ratings across survey administration modes for respondent subgroups: A MIMIC modeling approach. *Behavior Research Methods, 39,* 502–509.

CROSS-CUTTING ISSUES V

Anja S. Göritz

Using Lotteries, Loyalty Points, and Other Incentives to Increase Participant Response and Completion

14

ncentives are material and nonmaterial inducements and rewards that are offered to respondents in exchange for their participation in studies. This chapter explains the advantages and disadvantages of using incentives in Web-based studies and describes the types of incentives that are available. Moreover, the chapter seeks to develop evidence-based guidelines for short-term, as well as long-term, use of incentives to attain the goal of collecting high-quality data in a cost-conscious manner. Although a number of theoretical frameworks have been proposed to explain how incentives work (e.g., for an overview of theoretical accounts, see Singer, 2002), the focus of this chapter is pragmatic rather than theoretical.

By handing out incentives to respondents, researchers can increase the likelihood of people participating in Web-based studies, and incentives may improve the quality of respondents' responses. In particular, incentives can increase the response and the retention rates in a study. The *response rate* is the number of people who call up the first page of a study divided by the number of people who were invited or were aware of and eligible to take part in this study. The *retention rate* is the number of respondents who stay until the last page of a study relative to the number of respondents who have called up the first page of this study. Moreover, there is the hope—but not yet many data—that incentives will also increase other facets of data quality such as the completeness, consistency, and elaborateness of participants' answers.

However, using incentives might also entail undesirable effects. First, incentives might increase the response and retention rates at the expense of other facets of data quality, for example, more items are skipped; response styles occur more often; or answers to open-ended questions are shorter. It is possible that groups who are offered an incentive will answer less conscientiously than groups without incentives because the incentives might reduce the intrinsic motivation to perform the task (Heerwegh, 2006). However—at least in offline surveys—sometimes the opposite has been found to be the case (Singer, Van Hoewyk, & Maher, 2000). Singer (2002) found that "people who are rewarded for their participation would continue to give good information" (p. 168). The second potential undesirable effect of incentives is that they might attract a particular type of respondent and thereby bias sample composition (e.g., poorer people may be more responsive than richer people to monetary incentives; Groves & Peytcheva, 2008). The third possible effect is that incentives might actually reduce the response and retention rates by alienating intrinsically motivated volunteers (see Deci, 1971). Finally, incentives might bias the study results, for example, by altering the mood of the respondents (Singer, 2002) or by altering respondents' attitude to the researcher.

There is the risk that to earn an incentive, people with little motivation will fill in meaningless data to get to the end of a survey quickly. When no incentive is promised, bored people usually abandon the study prematurely, so they are easily identifiable. Moreover, when offering incentives, researchers need to follow ethical guidelines as well as legal regulations (see chap. 16, this volume). Because the laws pertaining to the use of incentives differ across some countries, particular care is necessary with international studies. Finally, in studies with ad hoc recruitment of respondents, incentives might induce some people to fill out and submit the questionnaire many times, and it is not always possible to detect skillful fraud. To weigh whether incentives can be recommended despite possible drawbacks, researchers need to know how large the desirable and undesirable effects are.

Implementation of the Method

In view of the already high and steadily growing importance of Internet-based data collection for both academic and commercial purposes, it is vital to develop methods for promoting the quality of gathered data. By handing out incentives for participating in a study, researchers can influence people's likelihood of taking part in studies and the quality of their responses. Besides incentives, other response-enhancing methods include prenotification, assurances of anonymity, stating a deadline, social-utility and help-the-researcher appeals, and reminders. However, meta-analyses routinely show that at least in traditional survey modes, incentives have a larger impact than other response-enhancing methods (Edwards et al., 2002; Yu & Cooper, 1983). Consequently, in the relatively new field of online data collection, the study of incentives has been given priority over the study of other response-enhancing methods. As a result, researchers can draw on a considerable body of knowledge on the effectiveness of incentives. However, to derive best practice guidelines, their context of use—that is, the type of study in which incentives are used, the sample to whom they are offered, and the character of the incentive itself—needs to be taken into account.

TYPES OF STUDIES AND SAMPLES

The sponsorship of a study, its length, and the saliency of the study topic to the target group are possible moderators of an incentive's effectiveness. There are reasons to assume that a material incentive is more effective if a study is commercially sponsored, if it is a long study, and if the study topic is less salient to the target group. In these cases, an attractive incentive may compensate for participation in an otherwise burdensome study (for a more detailed discussion, see Marcus et al., 2007).

The impact of incentives may also depend on whether participants are from an ad hoc or a list-based sample. *Ad hoc samples* are newly recruited from a pool of unknown individuals for a particular study. By contrast, list-based samples (e.g., drawn from the student body of a university, the staff of a company, or panelists from an online panel) contain identifying information for each list member. This allows potential respondents to be directly approached and invited to take part in the study. One widespread form of such sampling lists is an online panel. An *online panel* is a group of people who have agreed to occasionally participate in Web-based studies; thus, it is a pool of readily available participants for different kinds of studies—both longitudinal and cross-sectional (Göritz, 2007).

The decision-making dynamics of participants in online panels as to whether to take a part in a study may differ from ad hoc recruited people in several ways. First, unlike ad hoc recruits, panelists have already committed themselves to participating in a series of studies. Thus, a study invitation does not come unexpectedly, and a certain degree of trust has already been established. Consequently, an incentive is not needed to establish their willingness to participate but, rather, to increase interest in the study at hand. Next, compared with ad hoc recruits, who make quick decisions about participating when encountering banners or other announcements during Web surfing, panelists can more thoughtfully contemplate participation when they receive an e-mail invitation. Third, because each panelist's participation history can be used to decide who is invited to future studies, panelists might think twice about whether to decline an invitation. For one-time respondents, this is irrelevant. Finally, curiosity as a natural motivator is higher in ad hoc studies than in continuing panels, probably rendering external incentives less necessary. This effect might play out more severely the higher the frequency of studies in an online panel.

TYPES OF INCENTIVES AND THEIR LOGISTICS

The timing of incentives influences their effectiveness. An incentive can either be prepaid (i.e., awarded in advance) or promised (i.e., paid on return of the questionnaire). For the offline realm, researchers widely agree that prepaid incentives are more effective than promised incentives (Edwards et al., 2002) and even that prepaid incentives alone are useful (Church, 1993).

In the online world, however, using prepaid incentives is more difficult. Anonymous respondents in ad hoc studies cannot be sent cash or a physical token of appreciation. In this case, researchers need to make do with impersonal prepaid incentives such as free access to electronic information (e.g., e-books, e-journals, games, software). However, these prepaid electronic incentives might call forth a bias toward more computer-literate respondents, and it is doubtful that prepaid electronic devices are equally appealing to everybody.

Because in list-based studies people are personally addressable, delivering a prepaid incentive is easier than in ad hoc studies, and thus the array of feasible incentives is more extensive. If researchers know list members' e-mail as well as postal addresses, they have the choice between an incentive delivered electronically and one shipped by letter or parcel. Naturally, delivery by e-mail is cheaper. For example, one might include electronic gift certificates (i.e., vouchers). Researchers can buy electronic gift certificates in advance from many online stores and e-mail them to participants in due course. Researchers may get a discount if they purchase gift certificates in bulk. However, researchers should see to it that the gift certificates in question will not expire too soon. Other than electronic gift certificates, researchers can send redeemable loyalty points through e-mail. Loyalty points are a research organization's proprietary placeholder currency. A participant can exchange these points for actual currency, goods, or services—but usually only after he or she has collected a certain minimum number of points. Furthermore, electronic money (e-money) may be sent by e-mail using intermediaries such as PayPal. An option called a *"mass payment"* makes it possible to pay a large number of recipients simultaneously. To accomplish a mass payment, researchers compile a text file that contains the e-mail addresses with which the payees are registered with the intermediary along with the sum and currency to be paid accompanied by an optional "thank you" note. Researchers then upload this file onto the intermediary's server and effect the payment by pressing a button. However, e-money is not "money in the hand" (Bosnjak & Tuten, 2003), and depending on a respondent's Web literacy and willingness to register with a transaction party, collecting e-money is more or less cumbersome. Therefore, as an alternative, all participants, or only those who do not wish to register with an intermediary, can be asked their bank details at the end of the study in question or right when they register with a panel, and their money can be transferred to their own bank account. Upon request, many banks accept electronic files of pay orders to effect a bulk payment—similar to the process of mass payment described above. Of course, money, gift certificates, and redeemable loyalty points are not only suitable to be sent in advance but also upon the return of the questionnaire.

Indeed, the majority of Web-based studies appear to rely on promised incentives (Göritz, 2006a). This is probably due to the restricted choice and logistic challenges with prepaid incentives, as well as to the fact that promised incentives do work in the online realm (Bosnjak & Tuten, 2003; Göritz, 2006a). In contrast to offline studies, promised incentives work fine in Web-based studies probably because respondents are used to receiving incentives after their participation (Bosnjak & Tuten, 2003, p. 216).

In practice, the most common incentive used in online studies appears to be lotteries (Göritz, 2006a; Musch & Reips, 2000)—sometimes called *prize draws* or *sweepstakes*. A lottery costs the same regardless of how many people take part in the study. Although the prize that is raffled in a lottery is usually cash (Göritz, 2006a), it is equally possible to raffle something else, such as a physical gift, a gift certificate, or redeemable loyalty points. Raffling a few big prizes instead of several smaller prizes keeps transaction costs somewhat lower because fewer people need to be contacted and sent their prizes.

Possible incentives for Web-based studies other than lotteries, such as money, donations to charity, redeemable loyalty points, gifts, and survey results, have been used much more rarely (Göritz, 2006a). With the exception of survey results, the costs of such per capita rewards rise linearly with the number of participants. In the case of loyalty points, however, some of the paid points are never redeemed (e.g., this occurs if there is an expiration date for redeeming points or if list members discontinue their membership before redeeming their points). Furthermore, using loyalty points becomes more cost-effective if one's list management software supports automatic accounting of loyalty points (to read more about list management software, see the Additional Resources section). All of the mentioned incentives can also be used in combination—but so far hardly any research has examined the effectiveness of combined incentives.

THE OVERALL EFFECTIVENESS OF INCENTIVES

A meta-analysis of 32 incentive experiments (Göritz, 2006a) showed that material incentives increase response to Web-based studies; that is, incentives motivate people to start a Web-based study. Although significant, this effect is small: Incentives increase the odds of a person responding to a survey request by 19% relative to the odds without incentives. At the same time, material incentives promote retention: They increase the odds of a respondent to stay until the last page of a study rather than to drop out prematurely by 27% relative to the odds without incentives. The combined effect on response and retention for promised nonmonetary incentives in online environments—which are the type of incentives used most often—is larger than in offline environments (cf. Church, 1993; Edwards et al., 2002; Yu & Cooper, 1983).

Furthermore, there is no meta-analytic evidence that the incentive effects are dependent on whether the incentive is a lottery, prepaid versus promised, or monetary versus nonmonetary. Moreover, the incentive effect does not depend on whether the study is a nonprofit or commercial study, whether a result summary is offered in addition to the incentive, whether the sample is list based or ad hoc recruited, the total payout of the lottery, and the number of pages in the study. To sum up, it helps to use an incentive at all, but the type and amount of the incentive are less important. It remains to be seen whether these results stand up when more experiments become available.

STAND-ALONE EXPERIMENTS INVOLVING LOTTERIES OR LOYALTY POINTS

In an online panel survey (Göritz, 2004a), response with loyalty points was higher (82.4%) than when cash (78.0%) or trinkets (78.6%) were raffled. Loyalty points also brought about less dropout than the cash lottery, and the more loyalty points were awarded, the less dropout occurred. The total payout of the lottery as well as the splitting up of prizes in the lottery did not influence response, dropout, panelists' conscientiousness when filling out the questionnaire, survey outcome, or sample composition.

In Heerwegh's (2006) student survey, response and dropout were 63.5% and 7.8%, respectively, with a lottery of gift certificates and 60.2% and 10.3% without the incentive. Women were more responsive to the lottery than men. Students who were offered the lottery skipped fewer items than those in the control group.

In a study with ad hoc recruited participants, Marcus et al. (2007) found no difference in response and retention as a function of staging a lottery of gift certificates. In an online panel survey, Tuten, Galesic, and Bosnjak (2005) examined four levels of a cash lottery (i.e., high, medium, low, none) crossed with three levels of the timing of the notification of winners in the lottery (i.e., 12 weeks, 4 weeks, immediately). No linear effects were found for the amount of raffled money as well as the delay of notification on response and retention. However, in contrast to Heerwegh (2006), respondents with the lottery were more likely to skip items than control respondents.

In a market research online panel, Bosnjak and Wenzel (2005) compared (a) the promise of loyalty points, (b) prepayment of some loyalty points plus promise of further points, (c) a lottery of loyalty points with immediate drawing, and (d) a lottery of loyalty points in which the drawing was delayed for 1 week with a control group. With regard to respondents' conscientiousness, a stronger "yeah"-saying tendency was observed for the second option, prepayment of some loyalty points plus promise of further points. Response in this second option and the third option (a lottery of loyalty points with immediate drawing) and retention in the third option were higher than in the control group. This is an indication that respondents prefer to be immediately notified of whether they have won in a lottery—a phenomenon the authors termed the *immediacy effect*.

This immediacy effect was corroborated by Willige (2004), who contrasted several types of lotteries (i.e., cash, gift certificates, a holiday trip, or goods) crossed with different delays of drawing the prices (i.e., 3, 7, or 14 days) with a control group. Only when the most expensive prize (i.e., the trip) was raffled and drawn within 3 or 7 days was the response higher than in the control group. The immediacy effect was also corroborated by Tuten, Galesic, and Bosnjak (2004) but not by Tuten et al. (2005).

DONATIONS TO CHARITY

A donation to charity on behalf of a respondent in exchange for his or her participation could be an economical incentive because it incurs almost no transaction costs. However, when used in offline studies, donations tend to be ineffective (Hubbard & Little, 1988; Warriner, Goyder, Gjertsen, Hohner, & McSpurren, 1996). Göritz, Funke, and Paul (2009) contrasted the offer to donate 0.5 € or 1 € per respondent with the offer for inclusion in a 5 × 20 € lottery. Response and retention with the lottery were 46.1% and 93.7%, respectively; with the 0.5 € donation, 37.3% and 92.1%, respectively; and with the 1 € donation, 35.1% and 94.6%, respectively. In another experiment (Göritz et al., 2009), response and retention with a 1 € donation were 55.5% and 83.3%, respectively, and with a control group, 61.5% and 83.3%, respectively. However, before putting donations down as detrimental to the response rate, more experiments are needed.

STUDY RESULTS

If effective, study results are attractive incentives because they are cheap and might attract intrinsically motivated and therefore particularly conscientious participants. Survey results come in different forms. With regard to timing, there is continuous feedback while a participant is taking a survey, feedback immediately after having submitted the questionnaire, and delayed feedback after the data have undergone analysis. With regard to personalization, feedback can be tailored to the participant, and there is general feedback about the overall results of a study.

A meta-analysis from the offline realm yielded, in tendency, a negative influence of study results on response (Yu & Cooper, 1983). In the same vein, a review of several uncontrolled online panel studies shows result summaries to lower response but not to influence retention (Batinic & Moser, 2005). An online experiment (Göritz, 2009) yielded a response rate of 13.6% if a result summary was offered versus 15.2% without a summary. The dropout rate was 77.8% with results and 77.2% without results.

With an ad hoc sample for which the study topic was salient, Tuten et al. (2004) observed a response rate of 69.3% if results were offered versus 62.3% in the control group. Moreover, dropout was 40.7% with results versus 42.5% without. Finally, Marcus et al. (2007) offered either a personal feedback on individual results, general survey results, or no results at all, which yielded response and dropout rates of 34.6% and 19.7%, 33.1% and 22.2%, and 33.4% and 26.3%, respectively.

These inconclusive findings emphasize the need for further research on the usefulness of study results. As a tentative conclusion, however, offering results seems to be useful if participants are highly involved in the study topic or if the feedback is personalized. Moreover, study results seem to influence retention more positively than response.

CONDITIONAL INCENTIVES

Another simple and cost-free option is to mention in the request for participation that receiving an incentive is contingent on the submission of a completely filled out questionnaire. Restricting the eligibility for an incentive to people who did not skip any question might induce participants to fill out the questionnaire more completely and more conscientiously to be eligible for the incentive. Five experiments using cash lotteries or per capita gifts (Göritz, 2005) shed light on the effectiveness of this technique when used in online panels. Incentives that are contingent on the submission of a completely filled out questionnaire decreased response (odds ratio = 0.81) but did not influence dropout, the number of omitted closed-ended items, the length of answers to open-ended questions, and the tendency to answer grid questions in a stereotypical fashion. Thus, at least in online panels, it is not recommended to frame the incentive as being contingent on the completion of the questionnaire.

LONG-TERM EFFECT OF INCENTIVES

Because most of the hitherto presented evidence stems from cross-sectional studies, the degree to which these findings generalize to repeated surveys is not known. Knowledge on how to motivate people to participate in the long run is especially salient in the context of online panels because participants are requested to take part repeatedly.

In a five-wave experiment (Göritz, 2008), one group of panelists was sent a mouse pad as an advance gift; the other group did not receive a gift. For participation in the studies, half of the panelists were repeatedly offered redeemable loyalty points, and the others, inclusion in cash lotteries. At the outset of the series of studies, the advance gift significantly increased participation, whereby low-income panelists were more susceptible to the advance gift than were the well-off people. The gift was especially useful when combined with the lottery. Probably a tangible prepaid gift kindles trust, which extends to the lottery. However, the effects of the prepaid gift and its combination with the lottery faded throughout the waves of the study. Moreover, initially, there was no difference in participation between people with loyalty points and those offered to be included in the cash lottery; over time, however, loyalty points relative to the lotteries became more attractive.

In a four-wave experiment, Göritz and Wolff (2007) offered panelists repeated inclusion in a lottery of gift certificates or no incentive at all. Independent of the lottery, panelists who responded in a given wave were more likely to respond in the next wave. This process was characterized as a *Markov chain*. There was a direct positive effect of the lottery on response only at the first wave. However, mediated by the Markov process, the positive effect of the lottery on response at the first wave carried over into later waves. The lottery did not have any effect on dropout. Furthermore, it was found that dropout at a given wave is a reliable predictor for unit nonresponse (i.e., refusal to participate) in the next wave. Survey managers can use this information to diagnose and act on any impending nonresponse, perhaps by strengthening trust through personal communication, offering technical help, or a *refusal conversion incentive*, which is an extra incentive to persuade people at risk of dropping out to stay committed.

Six experiments conducted in a long-standing nonprofit online panel corroborated the ineffectiveness of lotteries in the long run (Göritz, 2006a). In each experiment, a cash lottery was offered in two versions:

either the total payout of the lottery was mentioned or the lottery was split into multiple prizes. The control group was not offered any incentive whatsoever. The lottery relative to no incentive did not increase response or decrease dropout; neither did it make a difference if one large prize or multiple smaller prizes were raffled. Panelists seem to extrapolate from their disappointing experience with previous lotteries that they are unlikely to win anything in the lottery at hand.

INCENTIVES FOR RECRUITING ONLINE PANELISTS

Müller-Peters, Kern, and Geißler (2001) showed that after face-to-face solicitation, people are more inclined to sign up with an online panel if cash, compared with no incentive, is offered or if a check versus inclusion in a lottery is offered. Furthermore, Göritz (2004b) examined whether recruitment was more successful if a cash lottery for new panelists was mentioned in appeals to join an online panel. The four solicitation methods for this study were random e-mail sampling from a volunteer Web-based sampling frame, random letter sampling, random fax sampling, and flier-based convenience sampling. The lottery was effective with fliers but not with e-mail, fax, or letter. Thus, when recruiting online panelists, material incentives can be useful, but only for certain solicitation methods.

Summing Up

When facing the choice of whether to use an incentive at all, it is advisable to offer an incentive, as material incentives demonstrably promote response and retention. When picking an incentive, its possible impact on response, retention, and other facets of data quality should be weighed against the cost of the incentive itself and its distribution in terms of money and manpower.

Regarding the type of incentive to be used, lotteries are usually cheap but may be ineffective. Lotteries are most useful in ad hoc studies and least useful when respondents have experienced the low likelihood of actually winning a prize such as in a longitudinal study or in a long-standing online panel. Based on the evidence so far, no straightforward relationship is discernible between the total payout or the number of prizes that are raffled in a lottery and the lottery's effectiveness. To save money, with a small research budget, it is recommended to keep the lottery payout to a minimum; with a bigger budget, it may be worthwhile to raffle one highly salient noncash prize such as a trip (Willige, 2004).

Moreover, respondents prefer to be notified whether they have won in a lottery as soon as possible. If the lottery at hand allows winners to be drawn in real time (e.g., in lotteries where every *n*th person wins a prize), this usually well-received fact should be communicated when soliciting respondents. Because with most lotteries, the expenses are independent of the number of participants, lotteries are especially economic—and thus recommended—if the expected sample size is medium to large. Donations to charity, if effective at all, seem to be less efficient than lotteries (Göritz et al., 2009).

If data are collected repeatedly from the same individuals such as in a longitudinal study or in an online panel, response and retention are higher with redeemable loyalty points than with lotteries. Also with loyalty points, so far no relationship has been found between the number of points offered and the response rate. However, as one might expect, dropout seems to decline with the number of points offered.

Because with loyalty points as with other per capita incentives, expenses rise more or less linearly with the number of participants, per capita incentives can become very expensive if the sample is large. To prevent costs from unexpectedly spiraling up, unless one can afford it, per capita incentives should be avoided in studies with ad hoc recruitment. It might happen that an unrestricted study gets unforeseen publicity with the consequence that many more people than expected do participate. Depending on one's budget, loyalty points and other per capita incentives are best used if the sample size is small to medium or if the participation of particular individuals is essential, as with low-incidence populations or in a longitudinal study.

If for some reasons lotteries are preferred even with repeated studies (e.g., for budget reasons or lack of a system for automatic accounting of loyalty points), the first lottery should be preceded by a tangible prepaid incentive (Göritz, 2008). Regardless of the type of incentive used, in online panels it is not recommended to frame an incentive as being contingent on the completion of the questionnaire (Göritz, 2005).

Besides being sparse in the first place, the evidence on how incentives influence sample composition and respondents' conscientiousness is mixed. Several of the reviewed studies have shown that some people are more susceptible to particular incentives than others (i.e., Göritz, 2008; Heerwegh, 2006), whereas other studies failed to find such an effect (i.e., Göritz, 2004a). However, it usually is a good idea to tailor the incentive to one's target group. The results pertaining to respondents' conscientiousness are also mixed: Whereas Tuten et al. (2005) and Bosnjak and Wenzel (2005) found that incentives had an impact on conscientiousness, Göritz (2004a) did not. These inconsistencies point to the presence of moderators (e.g., type of study, sample, incentive) that should be more closely examined in future studies.

TABLE 14.1

Effectiveness of Various Types of Incentives on Response, Retention, and Conscientiousness

Incentive	Response rate	Retention rate	Conscientiousness
Lottery	Ineffective to beneficial (Bosnjak & Wenzel, 2005; Göritz, 2004a, 2006a, 2006b,2008; Göritz & Wolff, 2007; Heerwegh, 2006; Marcus et al., 2007; Tuten et al., 2005; Willige, 2004)	Ineffective to beneficial (Bosnjak & Wenzel, 2005; Göritz, 2004a, 2006a, 2006b, 2008; Göritz & Wolff, 2007; Heerwegh, 2006; Marcus et al., 2007; Tuten et al., 2005)	Harmful to beneficial (Bosnjak & Wenzel, 2005; Göritz, 2004a; Heerwegh, 2006; Tuten et al., 2005)
Loyalty points	Beneficial (Bosnjak & Wenzel, 2005; Göritz, 2004a, 2006a, 2008)	Ineffective to beneficial (Bosnjak & Wenzel, 2005; Göritz, 2004a, 2006a, 2008)	Harmful to ineffective (?) (Bosnjak & Wenzel, 2005; Göritz, 2004a)
Donation to charity	Harmful to ineffective (?) (Göritz et al., 2009)	Ineffective (?) (Göritz et al., 2009)	?
Study results	Harmful to ineffective (unless personalized or highly salient) (Batinic & Moser, 2005; Göritz, in preparation; Marcus et al., 2007; Tuten et al., 2004)	Ineffective to beneficial (Batinic & Moser, 2005; Göritz, 2009; Marcus et al., 2007; Tuten et al., 2004)	?
Contingent	Harmful (?) (Göritz, 2005)	Ineffective (?) (Göritz, 2005)	Ineffective (?) (Göritz, 2005)

Note. The actual effectiveness may vary with type of study and sample. A question mark indicates that the outcome is preliminary because available data are scarce.

Table 14.1 gives a summary of the general effectiveness of different types of incentives on response, retention, and participants' conscientiousness. The statements about the effectiveness are based on—sometimes meager—evidence available to date, so they are tentative.

Outlook

Because the field of Internet-based studies is changing, with different segments of the population reaching the Internet and technological innovations coming up, the results of the present review are not final. Furthermore, previous experiments are insufficient in that they have thoroughly studied only a few types of incentives. Thus, comparatively comprehensive insights have been gained into the effectiveness of lotteries; some into redeemable loyalty points but few into the effects of donations to charity, per capita payments, and result summaries. Moreover, in most of the existing experiments, only a few facets of data quality were considered. Often, merely the response and retention rates were examined. However, comprehensive insights about trade-offs between different facets of data quality are needed. Finally, to get a more complete picture, the effects of incentives need to be examined under a wider array of circumstances with regard to study and sample characteristics.

Additional Resources

Meta-analysis of incentive effects: Readers are invited to submit their own experimental findings for inclusion in a future update of this review located on the supplementary site for this chapter.

List management software: The management of respondent lists in general and the allocation of incentives in particular are rendered more convenient if the researcher can avail of a Web-based list management tool. List management software is available from different companies. A Web search will bring up Web sites of relevant companies, and the reader can look up up-to-date information about the costs involved. Professional list management tools seem to be widespread among commercial online panels. By contrast, academic online panels mostly rely on homemade solutions. To my knowledge, only one freeware program exists for managing online panels and similar respondent lists. This free, open-source tool can be downloaded from the supplementary Web site for this chapter.

References

Batinic, B., & Moser, K. (2005). Determinanten der Rücklaufquoten in Online-Panels [Determinants of response rates in online panels]. *Zeitschrift für Medienpsychologie, 17,* 64–74.

Bosnjak, M., & Tuten, T. L. (2003). Prepaid and promised incentives in Web surveys—An experiment. *Social Science Computer Review, 21,* 208–217.

Bosnjak, M., & Wenzel, O. (2005, March). *Effects of two innovative techniques to apply incentives in online access panels.* Paper presented at the 7th Annual General Online Research Conference, Zurich, Switzerland.

Church, A. H. (1993). Estimating the effect of incentives on mail survey rates: A meta-analysis. *Public Opinion Quarterly, 57,* 62–79.

Deci, E. L. (1971). The effects of externally mediated rewards on intrinsic motivation. *Journal of Personality and Social Psychology, 18,* 105–115.

Edwards, P., Roberts, I., Clarke, M., DiGuiseppi, C., Pratap, S., Wentz, R., & Kwan, I. (2002). Increasing response rates to postal questionnaires: Systematic review. *British Medical Journal, 324,* 1183–1185.

Göritz, A. S. (2004a). The impact of material incentives on response quantity, response quality, sample composition, survey outcome, and cost in online access panels. *International Journal of Market Research, 46,* 327–345.

Göritz, A. S. (2004b). Recruitment for online access panels. *International Journal of Market Research, 46,* 411–425.

Göritz, A. S. (2005). Contingent versus unconditional incentives in WWW studies. *Advances in Methodology and Statistics, 2,* 1–14.

Göritz, A. S. (2006a). Cash lotteries as incentives in online panels. *Social Science Computer Review, 24,* 445–459.

Göritz, A. S. (2006b). Incentives in Web studies: Methodological issues and a review. *International Journal of Internet Science, 1,* 58–70.

Göritz, A. S. (2007). Using online panels in psychological research. In A. N. Joinson, K. Y. A. McKenna, T. Postmes, & U.-D. Reips (Eds.), *The Oxford handbook of Internet psychology* (pp. 473–485). Oxford, England: Oxford University Press.

Göritz, A. S. (2008). The long-term effect of material incentives on participation in online panels. *Field Methods, 20,* 211–225.

Göritz, A. S. (2009). *Survey Results as Incentives in Online Panels.* Unpublished manuscript, University of Würzburg, Würzburg, Germany.

Göritz, A. S., Funke, F., & Paul, K. (2009). *Donations to charity as incentives in online panels.* Unpublished manuscript, University of Würzburg, Würzburg, Germany.

Göritz, A. S., & Wolff, H.-G. (2007). Lotteries as incentives in longitudinal Web studies. *Social Science Computer Review, 25,* 99–110.

Groves, R. M., & Peytcheva, E. (2008). The impact of nonresponse rates on nonresponse bias: A meta-analysis. *Public Opinion Quarterly, 72,* 167–189.

Heerwegh, D. (2006). An investigation of the effect of lotteries on Web survey response rates. *Field Methods, 18,* 205–220.

Hubbard, R., & Little, E. L. (1988). Promised contributions to charity and mail survey response rates: Replication with extension. *Public Opinion Quarterly, 52,* 223–230.

Marcus, B., Bosnjak, M., Lindner, S., Pilischenko, S., & Schütz, A. (2007). Compensating for low topic interest and long surveys: A field experiment on nonresponse in Web surveys. *Social Science Computer Review, 25,* 372–383.

Müller-Peters, A., Kern, O., & Geißler, H. (2001, May). *Die Wirkungsweise unterschiedlicher Incentivierungssysteme auf Rekrutierungserfolg und Stichprobenqualität* [The effect of incentives on recruitment and sample quality]. Paper presented at the 4th Annual German Online Research Conference, Göttingen, Germany.

Musch, J., & Reips, U.-D. (2000). A brief history of Web experimenting. In M. H. Birnbaum (Ed.), *Psychological experiments on the Internet* (pp. 61–87). New York: Academic Press.

Singer, E. (2002). The use of incentives to reduce nonresponse in household surveys. In R. M. Groves, D. A. Dillman, J. L. Eltinge, & R. J. A. Little (Eds.), *Survey nonresponse* (pp. 163–177). Chichester, England: Wiley.

Singer, E., Van Hoewyk, J., & Maher, M. P. (2000). Experiments with incentives in telephone surveys. *Public Opinion Quarterly, 64,* 171–188.

Tuten, T. L., Galesic, M., & Bosnjak, M. (2004). Effects of immediate versus delayed notification of prize draw results on response behavior in Web surveys: An experiment. *Social Science Computer Review, 22,* 377–384.

Tuten, T. L, Galesic, M., & Bosnjak, M. (2005, March). *Optimizing prize values in Web surveys: Further examination of the immediacy effect.* Paper presented at the General Online Research Conference in Zurich.

Warriner, G. K., Goyder, J., Gjertsen, H., Hohner, P., & McSpurren, K. (1996). Charities, no; lotteries, no; cash, yes: Main effects and interactions in a Canadian incentives experiment. *Public Opinion Quarterly, 60,* 542–561.

Willige, J. (2004, March). *Vergabe von Incentives und Durchführung von Verlosungen: Maßnahmen zur Steigerung der Teilnahmebereitschaft in einem Online-Panel* [Incentives as measures for augmenting the willingness to participate in an online panel]. Poster session presented at the 7th Annual German Online Research Conference, Duisburg, Germany.

Yu, J., & Cooper, H. (1983). A quantitative review of research design effects on response rates to questionnaires. *Journal of Marketing Research, 20,* 36–44.

Olaf Thiele and Lars Kaczmirek

Security and Data Protection

Collection, Storage, and Feedback in Internet Research

15

This chapter discusses and provides recommendations about the various security issues associated with Internet research. The main topics are collection, transfer, and storage of data, and communication and feedback to respondents. An introductory summary of current legislation and ethical views enables researchers to make informed decisions about their infrastructure and research process. Also, this chapter alerts Internet researchers to dangers such as their data becoming invalidated or viewed by unauthorized persons, or having their results manipulated. It will be especially useful for people who maintain their own server or who store data on local networks and local computers. The checklists and explanations of technical terminology will enable researchers to communicate effectively with programmers or IT staff and help them to assess the security level of a research project or of proposed solutions.

The checklist in Exhibit 15.1 is a summary of what readers will learn in the coming chapter. It helps address the most important security issues in the different phases of research projects with examples in the languages PHP and the data-

We thank Mirta Galesic for allowing us to use the examples on informed consent.

EXHIBIT 15.1

Security Checklist

☑ Use HTTPS when handling personal data. Ensure a valid certificate and consider potential drawbacks.
☑ Ensure that the changing of URL variables by participants does not lead to unintended behavior.
☑ Choose sensible values for cookie names, expiration dates, and session timeouts.
☑ The software should check all input variables for special characters and length, especially in custom software.
☑ Limit physical access to the server and set appropriate access rights.
☑ Database users should have only limited rights depending on their role.
☑ Use HTTPS with authentication to download answers from the server; switch to SSH for sensitive and personalized data.
☑ Never send unencrypted personal data by e-mail.
☑ Delete and destroy personal data in the end.

base MySQL. Important security checks are highlighted throughout the text by summarizing check boxes, practical examples are presented in shaded boxes, and technical terms for the communication with IT staff are marked as Techspeak.

Legislation, Data Protection, and Ethical Issues

Legislation and professional ethics form the background for security and data protection issues. Data protection is based on the fundamental human right to privacy, which includes at least the following aspects: anonymity, confidentiality, informed consent, disclosure, and accuracy of data. Ethical issues are covered in depth in chapter 16, this volume. Online resources are also widely available (overviews are provided by Kaczmirek & Schulze, 2005; Madge, 2006). Data collected on the Internet can originate from many different countries, each having different data privacy laws. It is therefore advisable to draw on a data privacy approach that reaches beyond the home country's legislation when conducting online research (Ess & the Association of Internet Researchers, 2002). This is important because the United States and the European Union follow two different approaches to data protection. Whereas the United States implemented institutional review boards (IRBs; see chap. 16, this volume) and laws regulating different topics (e.g., the Children's Online Privacy Protection Act of 1998), the EU passed a general data protection directive (Directive 95/46/EC; European Parliament and Council, 1995), which has been enacted in each EU member country (e.g., the Data Protection Act 1998, Great Britain, 1998). The following demands of this EU directive are of special interest in this chapter: (a) Data should be accurate (Art. 6d); (b) identification of data subjects should not be possible longer than necessary (Art. 6e); (c) aspects of disclosure (Art. 11); (d) subject's right to access (Art. 12); (e) the right to object (Art. 14); and (f) security of processing (Art. 17). Article 17 is particularly relevant and states that researchers

> must implement appropriate technical and organizational
> measures to protect personal data against accidental or
> unlawful destruction or accidental loss, alteration, unauthorized
> disclosure or access, in particular where the processing involves
> the transmission of data over a network, and against all other
> unlawful forms of processing. Having regard to the state of
> the art and the cost of their implementation, such measures
> shall ensure a level of security appropriate to the risks
> represented by the processing and the nature of the data
> to be protected.

In most cases, informed consent is necessary before data collection is allowed to start (for possible exceptions, see Ess & the AoIR Ethics Working Committee, 2002). Figure 15.1 shows a text for informed consent prior to the start of online data collection. The following shaded box shows a shortened version that can be appropriate if respondents have given informed consent earlier (e.g., as a panel member). Both versions were approved by an IRB. As in Exhibit 15.1, it is important to inform respondents about the data protection measures beforehand.

Data protection is relevant when personal data are processed. *Personal data* are defined as "any information relating to an identified or identifiable natural person" (Art. 2a of Directive 95/46/EC). Individuals can be identified directly (e.g., by name) or indirectly by a number or other factors, which makes identification possible (see the Feedback to Participants section in this chapter). Many measures and features to enhance security are available as part of software products. Thus, off-the-shelf software products for Web-based data collection may already cover researchers' security and broader needs (Kaczmirek, 2008).

This introduction to legislation, data protection, and ethical issues forms the basis of the following security issues discussed below.

[Header stating the topic]

Welcome to the [Name of University].

We appreciate your cooperation. This survey should take about [estimated length] minutes. Your participation is voluntary and you may

FIGURE 15.1

Please fill out the following Consent Form and answer a few questions about yourself.

This study has been approved by the University ▨▨▨ Institutional Review Board. Information that you submit will be strictly confidential. No names or personal information will be used in any reports or publications. In order to continue, please complete the University ▨▨▨ Informed Consent Contract:

☐ I am 18 or over and have freely volunteered to participate in this project.

☐ I have been informed in advance as to what my tasks would be and what procedure would be followed.

☐ I understand that there are no known risks to my participation of this research, and that this research is not designed to help me personally.

☐ I am aware that I have the right to withdraw consent and discontinue participation at any time without being penalized in any way.

First name:		Last name:	

If you have questions about your rights as a research subject or wish to report a research-related injury, please contact:
Institutional Review Board Office, University ▨▨▨ (e-mail) ▨▨▨ (phone) ▨▨▨ -- You may contact the Chair of the Human Subjects Committee for any questions regarding the rights of a research participant.
Professor ▨▨▨ -- Principal Investigators: ▨▨▨ (e-mail) ▨▨▨ Phone: ▨▨▨

If you agree to participate in the study, please press Continue. Otherwise, please close this window.

Informed Consent With First Contact. From Web Survey conducted by Mirta Galesic. Copyright Mirta Galesic. Reprinted with permission.

skip any questions you prefer not to answer. All of your responses will be kept completely confidential.

If you have any questions or experience difficulty with the survey, you may contact us via e-mail at [e-mail address] or call toll free [phone number].

Should you have questions regarding your rights as a participant in research, please contact:

Institutional Review Board

[full contact information with name, address, phone number, e-mail]

Click NEXT to begin the survey.

Collecting Data From Respondents

This section deals with three security problems that arise when collecting data in behavioral research on the Internet. If participants are asked to reveal personal data in a questionnaire, they want to be sure that no unauthorized individuals are able to access their Internet connection. We therefore start by discussing data transmission between client and server. Second, we discuss typical problems and technical issues associated with ensuring that participants filling out a questionnaire are not able to open data sets other than their own. Finally, we present well-established countermeasures against malicious use of the software itself. These attacks are usually not executed by participants but by attackers for whom personal data are of monetary value (e.g., to send spam).

DATA TRANSMISSION BETWEEN CLIENT AND SERVER

Behavioral research frequently studies aspects of private life. Although participants might feel little reluctance to answer questions about food consumption, asking details about income is known to cause lower response rates. Most participants will have heard some kind of warning about unsafe transactions over the Internet, especially in the context of online purchasing, spam, and viruses. This should be taken into account when thinking about securing the connection. Participants should not be expected to enter personal data such as their address or ID numbers such as student number or Social Security number if no safe mechanism is offered. Similarly, they might refuse to download an executable encryption program (e. g., EXE file) for answering a questionnaire because they have heard that EXE files may contain a virus, although such a program might provide a high level of security. We therefore recommend using the built-in browser security mechanism HTTPS (hypertext transfer protocol secure), the most commonly used method for securing Internet connections between clients and servers. All common browsers support this protocol, and users do not need to download or pay to use it. Amazon and eBay, for example, use the protocol when switching to the log-in screen. It is easily recognizable by the start of the URL in the browser: A secured connection reads "https://" instead of "http://."

HTTPS is an extension to the standard HTTP (hypertext transfer protocol) used by browsers to present Web pages (Ristic, 2005). But HTTPS has two drawbacks that might rule out its use for certain research types: company policies and certificates. First, HTTPS uses a different port number for communication. The port number is an identification number that is needed because many protocols are running on a computer simultaneously. Corporate networks are often secured by a firewall that blocks most incoming and outgoing connections (ports) from and to the Internet. These measures are meant to prevent attacks from the outside, or in the case of universities, they keep inside users from downloading and sharing copyrighted material with the outside world. Only certain ports are forwarded in such an environment. This is usually Port 80 for the HTTP and Port 443 for HTTPS. Some companies block Port 443 to prevent employees from sending internal documents secretly or to keep them from using the Internet to, for example, shop and send private e-mails. Politically, some countries (e.g., China) forbid the use of certain strong encryption technologies. A comprehensive, regularly updated overview on local legislation is provided by Koops (2007). Second, most inexpensive Web servers do not offer trusted certificates. The main idea is that whenever two computers want to establish a secure connection, they need to trust each other. This is accomplished through a chain of trust. A top-level institution gives out certificates to subcertification agencies. Institutions or real persons then buy certificates from the agencies. All certificates stemming from the top-level institution can technically verify each other and can therefore establish a secure connection. Popular browsers (e.g., Firefox, Internet Explorer) are shipped with a built-in list of certificates issued by top-level institutions. When establishing a connection with one of the biggest verification institutions such as Verisign (https://www.verisign.com/), no warning message will pop up. Its noncommercial competitor CAcert (https://www.cacert.org/), on the other hand, is not included in the built-in lists, and thus a warning message pops up. In the past, this warning message simply informed users to think about the validity of the certificate, but with Version 7 of Internet Explorer and Version 3 of Firefox, the message reads, "We recommend that you close this Web page and do not continue to this Web site" and "Secure Connection Failed," respectively. This recommendation to abandon the Web site is likely to scare off potential visitors. Enabling access in Firefox 3 and above even requires a three-step process instead of a simple click. Universities and companies can circumvent the warning message within their realm by introducing a custom certificate to all their browsers (e.g., in all libraries). The warning will not appear on those computers, but it will appear on computers not belonging to the institution. Therefore, questionnaires using HTTPS should be tested once from an outside computer (e.g., at an Internet cafe) but not from a computer at home, which might already have the university certificate installed from a previous visit. The following shaded box shows how to find installed certificates.

Firefox distinguishes between built-in certificates and those installed by the user. To see installed certificates go to the *Tools* menu and click

Options. Click *Advanced,* choose the *Encryption Tab* and click *View Certificates. Authorities* are the built-in certificates, while the *Web sites* tab shows self-installed certificates. Check the accompanying information. The certificate should be valid for the period of the research.

Experience shows that it is advisable to offer both an encrypted and unencrypted version of a questionnaire if it is unclear where participants are recruited. In most questionnaires with security needs, the introductory page need not be encrypted, so all participants can access it. The link starting the questionnaire can point to an HTTPS encrypted Web page. Please keep in mind that HTTPS is a security mechanism that encrypts the connection but not the content. Because the content is not encrypted, the questionnaire itself does not need to be changed. Every questionnaire can be made secure through HTTPS by simply changing the Web address from "http://" to "https://," if the Web server is configured accordingly (see Ristic, 2005, for a step-by-step tutorial).

> ☑ Use HTTPS when handling personal data. Ensure a valid certificate and consider potential drawbacks.

A virtual private network (VPN; Scott, Wolfe, & Erwin, 1998) might be a possible alternative in some settings. VPN offers a higher level of security than HTTPS and is useful for companies to allow employees remote access while they are out of office. It is a software alternative that requires additional client and server software installation. In special cases such as attitude surveys, this would be an appropriate security measure to protect the data transmission between the server and the survey manager without any further effort. If the content itself needs to be encrypted in addition to the connection (e.g., for online banking), other mechanisms are available. Flash applications could be used to encrypt the content itself (Ludwig, 2005) in addition to the connection (e.g., with HTTPS). Nevertheless, the costs would exceed the benefit for most research applications.

IDENTIFYING PARTICIPANTS AND AVOIDING UNINTENDED ACCESS

Participants should not be able to view answers other than their own. This is often taken for granted, but online research sometimes proves the opposite. For example, early questionnaires stored the value for participant identification within the URL (Curphey et al., 2002) so earlier participants' answers could be accessed by simply changing the ID within the Web address to a smaller value (see the following shaded box). This should not work in current questionnaires, but testing URL variables for harmful consequences is advisable.

Regular URLs such as http://test.org/index.php can be used to send variables to a questionnaire. A question mark is appended to the URL, and variables are separated by ampersands. Adding the variable "page" with a value of 7 and the variable "user" with a value of 23 to the URL will yield http://test.org/index.php?page=7&user=23. A standard behavior for a software solution that uses such an approach would be to present the next Web page together with empty questions for User 23. Manually changing the user value to 12 would show the already filled out page 7 of User 12. Such a functionality must be avoided.

TechSpeak: Ask your developer what might happen if somebody manually changes the GET (as used in the exhibit above) and POST (not as easy to see) variables, especially with respect to so-called "hidden variables."

☑ Ensure that the changing of URL variables by participants does not lead to unintended behavior.

Nearly all the URL problems mentioned in this chapter can be solved by implementing so-called sessions (Wang & Katila, 2003). The idea of sessions is to connect the participant's client computer with the server using means other than an ID in the URL variables. Classic telephone lines connect exactly 2 participants for the duration of the phone call. Internet connections, in contrast, last only as long as it takes to load a single Web site. Therefore, it is most important to keep track of individual respondents with sessions. A session ensures that the server knows that a new connection comes from the same computer as the last one. If several users access the application at the same time, it keeps track of which requests are sent from which computer. Most modern Web applications use sessions. A session is usually initiated by entering a login name and a password.

To open up and keep a session alive, it is necessary to store some kind of information on the client computer. Values are typically stored on the client computer in the following way. The server tries to store a cookie, which is an alphanumeric value, on the client computer (see shaded box below and Wang & Katila, 2003). In a census of applicants to the University of Mannheim, we found that only 1.6% disallowed saving a cookie (Kaczmirek & Thiele, 2006). When a cookie is not allowed, the server appends an alphanumeric value (session ID) to every Web page sent to the client computer. Appending a session ID is only a more sophisticated technique than just sending the participant ID as mentioned above but is generally regarded as safer because of its length. Both cookies and session IDs can be altered by the user at any time. For behavioral research, this means that participants can end sessions at any time by deleting this information and can access the questionnaire as if it had never been started. Furthermore, if a participant can access another participant's browser, the former could theoretically write down the cookie value as described below and use it to access the latter participant's questionnaire. This should be kept in mind during data cleaning when encountering dubious multiple data entries that might be a result of data manipulation. Both problems can be avoided if the cookie is deleted by the software after questionnaire completion.

Some people see cookies as a variant of spam or as an intrusion into their privacy. It is therefore advisable to mention the use of cookies somewhere early in the introduction to the study. Cookies have an expiration date, which should be set to expire shortly after the data collection of the project finishes. The standard configuration for browsers is to delete a cookie on its expiration date.

To see which cookies are stored in your browser click in the "Tools" menu on "Options." Choose "Privacy" and click "View Cookies." Researchers should choose a self-explanatory name for the cookies with a reasonable expiration date. Unfortunately, names such as *uutz* and expirations in the year 2037 are common. Keep in mind that it is possible to configure the browser in a way so that all cookies are deleted each time the browser is closed.

TechSpeak: Ask your developer whether participants will be able to participate with disabled cookies. How is the problem solved if two different persons access the questionnaire from the same computer shortly after one another?

On shopping sites, sessions are usually initiated by entering a login name and a password, but questionnaires are a different case. Sometimes they may start seamlessly without the need for entering such identification. In other settings, participants are required to enter some sort of identification code, typically sent by e-mail; this raises a different class of problems: typing errors and wrong links. Sometimes users have problems copying long Web addresses from an e-mail into the browser. Also, Web addresses might be too long to be displayed in one line within the e-mail client so a click does not lead to the correct page. The best solution is to choose a short URL (less than 70 characters) that fits onto one line of an e-mail. The shaded box below describes a service that provides short links redirecting browsers to the origin.

If your URL cannot be shortened, tinyurl.com can be used. A typical URL from this free service looks such as http://tinyurl.com/3245yt/. This link can represent a URL of any length. Personalized e-mails can be sent by appending variables to the URL (e.g., http://tinyurl.com/3245yt/?ID=333, where 333 is the participants' identification).

Another advantage of sessions is their ability to store data within the session so a participant cannot change them. Therefore, information that would be problematic in URLs (e.g., the page number) should be stored within the session to avoid manipulation by participants. A further security measure is the session timeout value. If no contact from the client is made to the server for a set timeout period, the session is deleted. This value should be set to a sensible value. The timeout value could be 30 minutes, assuming that a single survey page or a break will not last longer even for the slowest participants.

☑ Choose sensible values for cookie names, expiration dates, and session timeouts.

MEASURES AGAINST CRIMINAL ATTACKS

People attacking the questionnaire and database may have monetary motives: E-mail addresses can be sold for advertising and spamming, and employee surveys contain private data valuable to competitors. Although opening unknown e-mail attachments on desktop computers can easily result in catching a virus, the analogous problem for online questionnaires is any form of input sent by the participant's browser. Open-question formats are designed to allow participants to answer a question, but malicious users can use them to introduce executable code into the software system (see the shaded box below). If successful, this code can change the behavior of the application or may give access to information that is otherwise unavailable from the database. The first case is called *code injection* the latter, *SQL injection.* Imagine a 360-degree feedback application that stores the user name within the session. When asking a usual "Anything else?" type of question, the open question could be used to inject code that switches the correct user name with a different one. Negative statements might then be used to slander other employees concealed under a different identity. Similarly, a criminal could insert SQL code that shows all previously saved e-mail addresses instead of the next page of the questionnaire. Closed-question formats are also vulnerable to attacks. Instead of returning a fixed value, rating scales can easily be altered to return malicious code. To prevent this sort of attack, all incoming data need to be checked. The standard procedure, which has proven to be effective, is to eliminate hazardous characters for the respective programming language (Peikari & Chuvakin, 2004; Ristic, 2005).

Typically, answer lengths for open questions are limited through HTML. The attribute "length" can for example be set to allow only 100 characters. But this does not prevent attackers from sending 500 characters or more to the questionnaire application. Therefore, the limits should be set both in HTML (so real participants are limited) and in the software (check length and content before storage). If the variable name is "answer5" and should contain only text, the following code example checks appropriately:

```
if (strlen($answer5) > 100) discardAnswer();
if (strpos($answer,";")) discardAnswer();
```

The first verifies that "answer5" is not longer than 100 characters, and the second checks whether "answer5" contains a semicolon (which is used to end a command and allows to insert malicious code). If such characters are needed for the answer, they can be encoded to avoid direct processing by the programming language.

Similarly, SQL statements might contain extra apostrophes that change the intended usage. The first line of code shows a way to save answers in the database. The second line shows an attack that would delete all results in the database.

```
mysql_query("UPDATE tableA
SET answer5=$answer5"); (storing data)
5'; DELETE * FROM tableA (attack by sending a changed
value for $answer5)
```

TechSpeak: Ask your developer whether all incoming variables are checked to be in the appropriate format, including hidden variables.

☑ The software should check all input variables for special characters and length, especially in custom software.

Storage of Data on Servers

Access to personal data is not restricted to the respondents alone. Many other people (e.g., researchers, research assistants, developers, system administrators) have contact with the data in part or as a whole in later stages. Therefore, the security levels in later phases of a project need to be as good as in the early stages. This section includes general guidelines for server security, recommendations for separating answers and personal attributes, advice on exchanging and deleting data, and recommendations for identifying attacks.

GENERAL GUIDELINES FOR SERVER SECURITY

The first rule is to restrict physical access to the server hosting the questionnaire. The simplest attack is copying the hard disk. A more common nuisance, though, is people switching off the server with the intent of saving energy. Putting up a sign stating that the server should not be turned off, together with the administrator's cell phone number has proven to prevent outages. The second rule is to restrict access to the software. Access rights on the computer should be set to exclude everybody not involved in the project. The two aspects of access (physical and software) are usually met by keeping the server in a special location with remote access (e.g., the university data center).

☑ Limit physical access to the server and set appropriate access rights.

Access rights to the database are a different matter because they could easily be set individually but typically are not. Users in databases are analogous to those in operating systems and can typically perform one of three operations: alter data (insert or update), extract data (select), and remove data (delete). The questionnaire application should only be able to insert or update data. Researchers should only be able to extract data partly, depending on the research objectives. Only administrators should have the right to delete data. If access rights are set up this way, then SQL injections (see, e.g., the shaded box above) would have close to no effect, as attackers could only insert fake data sets, but data extraction or deletion would be impossible.

TechSpeak: Ask your developer to implement user roles in the database. The questionnaire application should have the fewest possible rights. Researchers should have limited rights, too, if personal data are involved. Database views offer the possibility to show tables only in part and might be helpful. Deletion rights are dangerous; they should be granted sparingly.

☑ Database users should have only limited rights depending on their role.

SEPARATION OF ANSWERS AND PERSONAL DATA

By law, answers and personal data must be separated into different places. As the data are usually stored in a relational database (e.g., MySQL), the different places should be tables. The answers should be stored in one table, and personal data should be stored in a different table. Some questionnaires need documentation on who took part only for verification purposes, so personal data are not used. In this case, there should be no connection between the database tables. If a connection between answers and personal data are needed, it might be possible to use a one-way function (see the shaded box below) to meet legislative demands. A one-way function instantly transforms personal data (e.g., surname, date of birth) into an alphanumeric string, which would take months to decrypt. One-way functions are therefore suitable in situations in which connecting answers and personal data should be possible for those knowing the personal data (e.g., data protection officer) but not for those analyzing the answers (e.g., research assistants).

PHP offers the possibility of using hash functions. *Hashes* are one-way functions needed frequently in cryptography and are easy to use for data protection purposes. Typically, the e-mail address is inserted into a hash function, and the resulting value is used for identification of the respondent. The e-mail address as such is never saved. The following transforms an e-mail address to a hash value:

```
$mail_hash = hash('sha1', 'thiele@uni-mannheim.de' +
$secret_password);
```

This would result in the respondent ID 53b2499bae3364940e1879 cf1cd67583246ce56d. The "secret_password" variable is only known to the data protection officer. This way, no other personnel are able to connect personal data with answers.

One intuitive approach is to use randomly generated codes for anonymous data storage, but this method identifies returning respondents as new cases. Hash functions, however, return the same code given the same input, thereby allowing returning respondents to be identified; thus, hash functions are a good way to identify participants in longitudinal questionnaires without ever revealing their identity in a data set. They also provide a fast, easy, and secure solution to participants' requests for data deletion or information.

SECURE DOWNLOAD OF THE DATA SET

We have already covered safe collection of data on the Internet, but the subsequent download and exchange of data sets need a similar security effort. Most data protection failures occur in these later phases. For example, sending unencrypted personal data by e-mail might reach the wrong recipient because of a typing error.

In the beginning of the collection phase, researchers update the local data set frequently to check data patterns and to ensure that everything is going well. Typically, the database running on the Internet server is accessed through a Web interface (e.g., phpMyAdmin). This directory is frequently secured with a login screen that pops up when loading the URL. Technically, this HTTP authentication (Ristic, 2005) is implemented through an htaccess file for the Apache HTTP Server (Microsoft's IIS works analogously with dialogues. Depending on the researcher's needs, two different authentication mechanisms can be applied: by IP address and by user. The IP address can easily be forged and should therefore be used only as an authentication mechanism for testing purposes. Authentication by user is superior, but attacks are still possible. Attackers can force their entry by trying every possible password combination (i.e., *brute force attack*). Blocking access to a directory on an Internet server with an htaccess file is not advisable when handling personal data but works well for noncritical questionnaires, especially in combination with HTTPS to secure the connection.

When dealing with personal data, more secure techniques to download the data set should be applied instead of using HTTPS with authentication. Secure Shell (SSH) is one option that comes with several advantages. For example, access to the downloads might easily be restricted to a few computers. However, users need a special software client. In situations in which both the server and the local computer are nearby and behind a firewall, standard protocols to access databases (e.g., Open Database Connectivity [ODBC]) are acceptable, too. ODBC is an interesting alternative in low-security environments for various reasons: It is flexible, offers numerous options to select data from the data set, and is supported in standard office applications.

☑ Use HTTPS with authentication to download answers from the server; switch to SSH for personal data.

STORAGE AND EXCHANGE OF THE DATA SET

After the data are downloaded, they are stored on the local computer. Regarding personal data, many data protection laws demand that the critical information is stored safely. We recommend storing the data in few places with restricted access. The software TrueCrypt (http://www. truecrypt.org), for example, is an open-source product that stores critical data safely on both Windows and Linux. Even USB flash drives can be encrypted. Even though we advise storing personal data in only a few places, we recommend frequent backups (e.g., on CD) to prevent data loss. Software such as TrueCrypt can be applied to backups without difficulty.

Data, while being analyzed, are frequently exchanged within a workgroup or with outside researchers. In such cases, personal data need to be stripped from the results. If this approach is not feasible, the data should be sent securely to the recipient. We recommend two different techniques, depending on the situation. If e-mail communication is frequent, we suggest using PGP (Pretty Good Privacy) together with the e-mail client. The Thunderbird e-mail client can be extended by the Enigmail (http://enigmail.mozdev.org/) extension to send and receive encrypted e-mails. PGP uses a chain of trust similar to HTTPS as described previously. The initial effort of getting certificates and installation of the software is high, but e-mails will be encrypted and decrypted automatically whenever possible, ensuring the maximum available security in e-mail communication. If PGP seems too cumbersome, we recommend tools that encrypt the files sent, not the communication as a whole. LockNote (http://locknote.steganos.com/) is a simple software that offers strong encryption methods, can be installed in a few minutes, and encrypts and decrypts files using only a password.

☑ Never send unencrypted personal data by e-mail.

FINAL DELETION OF THE DATA

All personal data should be deleted as early as possible. This includes erasing data from the hard disk as well as destroying backup CDs and printouts. It is not sufficient to click "Delete" on a file to obliterate information on a hard disk because many programs are available to restore such perfunctory deleted data (Gutmann, 1996). Instead, we recommend using special software. The software Eraser (http://www.heidi.ie/eraser/) implements effective methods (e.g., defined by the U.S. Department of Defense) to delete files on both hard disks and other storage devices (e.g., USB flash drives). CDs and DVDs are best destroyed by scraping off parts of the reflective layer (the top side) or by cutting them into pieces.

☑ Delete and destroy personal data in the end.

IDENTIFYING ATTACKS

Professional intruders know how to conceal their attacks and will not leave any traces behind. Therefore, the first countermeasure should be to embed detection into software to find out whether somebody has indeed tried to intrude an application. The following advice should be taken into account when developing custom software. At the login phase, applications could deny access for half an hour in cases in which the wrong password is given three times in a row. This step deters criminals from quickly trying every possible password combination (i.e., brute force attack). Many shopping sites have implemented this safety feature, but a number of online questionnaire solutions do not block such attempts. In addition to blocking access, the IP address of the client should be stored in such situations because it might be used to identify attackers. Technically, a small script could send an e-mail notification each time such lockouts exceed a certain threshold. Such countermeasures clearly enhance the security level.

> *TechSpeak:* Ask your developer what is done to detect criminal intrusion and how such an intrusion is communicated to the research group (e.g., through e-mail).

Feedback to Participants

When data collection and storage are finished, researchers move on to aspects of analysis and reporting. In this section, we first cover how individual feedback can be given to participants. Second, we describe feedback procedures for aggregated data. These can be distributed to participants after analyses are complete or immediately after the participants have provided their data; we deal with both forms of feedback, each of which has its own special applications and poses different challenges to data protection.

SINGLE RESPONDENT FEEDBACK TO AN INDIVIDUAL

When participants receive feedback on how they scored, the feedback is targeted at an identified natural person and includes personally relevant data. Sometimes participants need to save their answers for their own documentation (e.g., from an establishment survey), in which case the feedback should contain all the questions and the participants' answers. For such sensitive material, special care is needed to prevent other persons gaining access.

Data collection can be classified as either personal or anonymous participation. In personal participation, e-mail invitations may be sent to respondents containing a password or link that allows them to retrieve the feedback repeatedly from a Web site. In anonymous participation, feedback needs to be administered on the pages following the questionnaire. After the feedback window is finally closed, respondents should be unable to regain access to this information, so they should be warned to save the page (see shaded box below). A possible workaround (e.g., with cookies) is not suitable because different people may have access to the same browser (e.g., at a university). Furthermore, cookies are prone to manipulation.

> *Note.* The report sent to your computer screen upon the completion of the IPIP-NEO is only a temporary Web page. When you exit your Web browser, you will not be able to return to this URL to re-access your report. No copies of the report are sent to anyone. If you want a permanent copy of the report, you must save the Web page to your hard drive or a diskette and/or print the report while you are still viewing it in your Web browser. If you choose to save your report, naming it with an .htm extension (e.g., Myreport.htm) as you save it may help you to read it into a Web browser later. (Johnson, n.d.)

FEEDBACK OF MULTIPLE RESPONDENTS TO INDIVIDUALS OR GROUPS OF PEOPLE

Feedback can be composed of aggregated data to ensure anonymity or to summarize different respondents' answers. If removal of personal data is done properly, the removal should be complete and thus data protection rules no longer apply. In open Web polls, for example, the last page often displays simple frequency or percentile distributions summarizing earlier respondents' answers (e.g., http://www.demandi.co.uk/). Chen, Hall, and Johns (2004) took an ethical perspective and recommended returning results of research in exchange for the participants' effort and answers. The results may be accessible by all Web visitors, if three conditions are met:

1. Respondents are not identifiable during participation or invitation;
2. open question formats do not reveal the identity of respondents; and
3. the number of respondents is high enough to prevent identification by "good guessing."

The first condition is easily met if no personal data are necessary or saved during participation. The second condition makes it necessary to postprocess open-ended questions manually (if they are included) and to mask information that may lead to identification of participants. Open answer fields need to be cleaned of names and contact information. The third condition is relevant when the sample frame is known or small subsamples are used. Employee satisfaction surveys are an example with a known sample frame because the eligible people are obvious. In the case of a small subsample, imagine a small department consisting of just two employees; here, analyses should be aggregated with the next higher level of the organizational structure to protect the privacy of the two respondents.

Finally, when using e-mails, researchers should send feedback reports separately to avoid identification of coparticipants. This can easily be done automatically by a PHP script using the mail function. If no scripting approach can be used and single e-mails are not an option, the blind carbon copy ("bcc" field) should be used for all addresses.

In this chapter, we provided both the legislative background and the technical advice on how to deal with security in behavioral research on the Internet. Using the summarized checklist in Exhibit 15.1 will allow one to concentrate on research while addressing necessary security aspects.

Additional Resources

First, please visit the supplementary Web site (http://www.apa.org/books/resources/gosling) for this chapter. The section for this chapter makes it easy to follow all mentioned Internet resources and links. It also contains the compiled checklist for easy printout.

GENERAL SECURITY ADVICE

Bruce Schneier Blog and Newsletter: http://www.schneier.com. Schneier has written a well-known cryptography textbook (*Applied Cryptography* [Schneier, B., 1996]) and offers an excellent blog and newsletter. The *Crypto-Gram* newsletter has more than 125,000 readers and regularly informs them about security related problems.

PHP AND MYSQL DOCUMENTATION

Both PHP and MySQL offer important documentation on available functions and commands. Each Web site has a section where it is possible to add comments to the official documentation, which usually gives good advice on security and other issues. The PHP Web site (http://www.php.net/manual/en/) offers a good section on overall PHP security. The MySQL documentation (http://mysql.org/doc/) presents a shorter section on security but generally informs well on database subjects.

MAILING LISTS ON CURRENT VULNERABILITIES

Local CERTs (Computer Emergency Readiness Teams) offer informative official mailing lists on vulnerabilities. The US-CERT (http://www.kb.cert.org/vuls/) offers constantly updated technical information (approximately once a day). Other CERTs can be found on the CERT Web site (http://www.cert.org/). The BugTraq (http://www.securityfocus.com/) mailing list has been around for years and sends unfiltered, high-volume information on security vulnerabilities.

References

Chen, S. S., Hall, G. J., & Johns, M. D. (2004). Research paparazzi in cyberspace: The voices of the researched. In M. D. Johns, S. S. Chen, & G. J. Hall (Eds.). (2004). *Online social research: Methods, issues, and ethics* (pp. 157–175). New York: Peter Lang.

Curphey, M., Endler, D., Hau, W., Taylor, S., Smith, T., Russell, A., et al. (2002). *A guide to building secure Web applications: The open Web application security project.* Retrieved April 28, 2007, from http://www.cgisecurity.com/owasp/html/

Ess, C., & the AoIR Ethics Working Committee. (2002). *Ethical decision making and Internet research: Recommendations from the AoIR Ethics Working Committee.* Retrieved March 30, 2007, from www.aoir.org/reports/ethics.pdf

European Parliament and Council. (1995). *Directive 95/46/EC on the protection of individuals with regard to the processing of personal data and on the free movement of such data.* Retrieved March 30, 2007, from http://eur-lex.europa.eu/LexUriServ/LexUriServ.do?uri=CELEX:31995L0046: EN:HTML

Great Britain (1998). Data Protection Act 1998. London: The Stationary Office. Retrieved July 27, 2009, from http://www.opsi.gov.uk/ ACTS/acts1998/19980029.htm

Gutmann, P. (1996). Secure deletion of data from magnetic and solid-state memory. *Proceedings of the 6th Annual USENIX Security Symposium* (pp. 77–90). San Jose, CA: USENIX Association. Retrieved May 5, 2007, from http://www.usenix.org/publications/library/proceedings/ sec96/full_papers/gutmann/

Johnson, J. A. (n.d.). *IPIP-NEO narrative report.* Retrieved June 5, 2007, from http://www.personal.psu.edu/faculty/j/5/j5j/IPIP/ipipneo samplereport.html

Kaczmirek, L. (2008). Internet survey software tools. In N. Fielding, R. Lee, & G. Blank (Eds.), *Handbook of online research methods* (pp. 236–254). London: Sage.

Kaczmirek, L., & Schulze, N. (2005). *Standards in online surveys: Sources for professional codes of conduct, ethical guidelines, and quality of online surveys.* Retrieved May 4, 2007, from http://www.Websm.org/uploadi/editor/ 1133803871kaczmirek-schulze2005-standards.pdf

Kaczmirek, L., & Thiele, O. (2006, March). *Flash, JavaScript, or PHP? Comparing the availability of technical equipment among university applicants.* Poster session presented at the General Online Research 2006 (GOR06), Bielefeld, Germany.

Koops, B.-J. (2007, January). *Crypto law survey.* Retrieved April 28, 2007, from http://rechten.uvt.nl/koops/cryptolaw/

Ludwig, A. (2005). *Macromedia Flash Player 8 security (white paper).* Retrieved June 5, 2007, from http://www.adobe.com/devnet/flash player/articles/flash_player_8_security.pdf

Madge, C. (2006). *Online research ethics.* Retrieved March 30, 2007, from http://www.geog.le.ac.uk/orm/ethics/ethcontents.htm

Peikari, C., & Chuvakin, A. (2004). *Security warrior.* Cambridge, England: O'Reilly.

Ristic, I. (2005). *Apache security.* Cambridge, England: O'Reilly.

Scott, C., Wolfe, P., & Erwin, M. (1998). *Virtual private networks* (2nd ed.). Cambridge, England: O'Reilly.

United States (1998). Children's Online Privacy Protection Act of 1998. Washington: U.S. G.P.O. Retrieved July 27,2009, from http://www. ftc.gov/ogc/coppa1.htm

Wang, P., & Katila, S. (2003). *An introduction to Web design and programming.* Pacific Grove, CA: Thomson Learning.

Tom Buchanan and John E. Williams

Ethical Issues in Psychological Research on the Internet

16

> With great power, comes great responsibility
>
> *—Spider-Man, 2002*

Since the mid 1990s, growing numbers of psychologists have become aware of the potential the Internet has for pure and applied research of various sorts. Discovering this rich and seemingly endless source of participants to recruit and data to mine, and a versatile new environment in which to conduct research, has been a truly empowering experience. However, along with the new opportunities have come new responsibilities, to participants and colleagues, for the ethical conduct of research.

The recognition that the Internet may have characteristics that require consideration over and above traditional techniques, as well as the development of guidelines for good practice, has taken time. Although there has been discussion of ethical issues ever since the discipline of Internet-mediated research began to emerge, it is only relatively recently that organizations such as the American Psychological Association (Kraut et al., 2004), the British Psychological Society (O'Dell et al., 2007), and the Association of Internet Researchers (2002) have published guidelines for the ethical conduct of online research.

This chapter is intended to provide a summary of some of the key ethical issues that online researchers need to consider. First, we consider what is special about online research environments in terms of ethical considerations and why there might be ethical constraints over and above those encountered in traditional investigative paradigms. Second, we consider some of the special concerns that may arise for different types of Internet-mediated research (e.g., questionnaire-based assessments, qualitative work). Third, we present some practical advice on satisfying ethical review committees, such as institutional review boards (IRBs), that determine whether project proposals meet ethical requirements. Finally, we present a selection of relevant publications and resources.

The chapter should be useful to anyone engaged in, or contemplating, any form of research that uses the Internet as a source of participants (e.g., recruitment of special groups for questionnaire-based research), a source of naturally occurring data for secondary analyses (e.g., field observations, analyses of archived discourse), or an environment in which the research process occurs (e.g., interactive experiments, online interviews). It should enable readers to decide whether particular research designs are appropriate for implementation over the Internet, and if so, what special procedures might need to be put in place. For example, is it appropriate to show excerpts of violent video clips online and measure changes in viewers' moods? What problems might arise if a researcher published analyses of material disclosed publicly in an online support group? It will also be of interest to editors and reviewers of journals to which online research is submitted for publication and may be of value to members of IRBs in helping them to decide whether research proposals have satisfactorily addressed the issues we describe. Ultimately, as the use of the Internet for psychological research increases, all producers and users of psychological research will need to be aware of the issues we discuss.

What Is Different About the Internet?

An obvious question to begin with is why psychological research conducted through the Internet should be any different from any other form of research. In what ways are the ethical implications different? In general, researchers tend to have considerably less control over, and knowledge of, the research environment and experience of participants than in lab-based studies and may have limited knowledge of who their participants are. This has implications for key components of ethical practice (e.g., gaining informed consent, debriefing participants, detecting and preventing harmful effects of participation). However, one must recognize that not all forms of online research are the same: There is considerable diversity in the techniques available, as glancing at the table of contents of this volume will confirm.

One also needs to recognize that there are different ethical frameworks and different points of view on what is acceptable in Internet-mediated research of various sorts. For example, one key issue is whether a piece of research involves "human subjects." The notion of "human subjects" of behavioral research and the ethical principles that must inform this research (e.g. respect for persons as autonomous agents, beneficence, and justice) are at the core of U.S. legislation on ethical standards and the network of IRBs whose role it is to enforce it (Code of Federal Regulations; U.S. Department of Health and Human Services, 2001). In research defined as involving human subjects, investigators obtain data directly through interactions with individuals. Research in which there is no interaction between the parties, but in which investigators access private information about identifiable individuals, is also classified as human subjects research.

The extent to which various types of Internet-mediated research can be classified as involving human subjects is somewhat open to debate. Much of the debate centers on notions of what constitutes private information and what is public. For example, is an entry on a blog that reveals sensitive and personal information to be considered private or public? What is the person's expectation of privacy: Does he or she consider this as akin to a private diary or to a published autobiography? Is the individual who posted it identifiable, either by his or her real name or a pseudonym (or *handle,* a screen name or pseudonym typically adopted online)? Does association of the material with a pseudonym count as identifiability? What if the posting is not on a blog but in an Internet-mediated support group of some kind? Do the same rules then apply? Kraut et al. (2004) presented a flowchart illustrating the key decision points as to whether a research procedure should be considered as involving human subjects and, therefore, requiring informed consent and coming under the jurisdiction of IRBs. In their discussion, they pointed to a number of gray areas that the Internet introduces to such decisions. Some of these gray areas are evident among the ethical considerations discussed next, which are listed in no particular order and overlap somewhat.

1. INFORMED CONSENT

The key issues here are whether it is necessary to obtain informed consent, and if so, how it should be obtained and documented. Kraut et al.'s (2004) flowchart can help researchers make these decisions. Skitka and Sargis (2006) suggested that most IRBs are concluding that "online postings represent the public domain and researchers do not need to obtain informed consent to use this material" (p. 549). However, it must be recognized that there is debate among researchers on this topic, especially in the case of online postings for which the poster's perception of privacy may be greater than is actually the case.

2. TRACEABILITY

The extent to which online research participants are identifiable or traceable may vary. In some cases, such as analyses of postings in a discussion group, it may be relatively easy to identify who a person is. The same is true of situations in which participants provide data directly to researchers (e.g., survey research) and some form of identity verification mechanism is in place. In other instances, though, participants may be entirely anonymous and untraceable by the researchers. This becomes problematic when one considers retrospective withdrawal of data, which some ethical codes (e.g., that of the British Psychological Society) require to be possible. If one gathers information from which identities can be verified, then other problems arise—most notably, keeping that personal information secure. One possible solution to some of these problems is to anonymize data but give participants unique ID codes (also stored in the data set), perhaps shown on a debriefing page. Presentation of these codes could enable individuals' data to be identified should they subsequently wish to withdraw it.

3. AUTHORS VERSUS SUBJECTS

Among others, Bruckman (2002) discussed the notion that when people post material on the Internet, they could be seen as adopting the role of amateur artists publishing their work, rather than potential human subjects of research. Hence, when analyzing postings in newsgroups or blog entries, for example, or including quotes from postings in articles, researchers should perhaps be more concerned with issues such as intellectual property and copyright than with issues such as informed consent. Different academic disciplines may have different views: Whereas the position adopted by psychologists would be that people's identities should be protected, researchers in other disciplines could make an equally strong case that people need to be credited as authors of their creative work. The boundaries of what is considered "published" versus "unpublished" (e.g., whether a blogger is an author or a diarist) become blurred online.

4. TECHNICAL ISSUES: INCOMPLETE PARTICIPATION

When Internet communication technology works, it is great. But servers crash, links drop, and various other problems occur. This means that interactive procedures (e.g., filling in a questionnaire, submitting it, and receiving a debriefing page) can sometimes be interrupted, and participants may be unable to complete a study. In a laboratory study, if something goes wrong, the experimenter will be there to explain, debrief, reschedule, and so on. However, in the online situation, the participant may just be left sitting alone at his or her computer staring at a blank screen or error message. This has consequences for the researcher (loss of data) that are annoying but not too serious (there may be plenty more participants "out there," so he or she can afford to lose a few). What may be more serious are consequences to the participant. If the procedure is not completed, he or she may not see a debriefing page: Thus, not only is he or she left unthanked and wondering what has happened but he or she also may not see information that is important for ethical reasons. Such information might include contact details for the research team, information on retrospective withdrawal of data, and full debriefing information. This last item is especially important in cases in which the study might have had an effect on the participant (e.g., a mood manipulation, something potentially distressing) or in which deception has been used.

5. DROPOUT

The same is true of situations in which participants decide not to complete an interactive study and drop out at some point. In a face-to-face (FTF) setting, it is actually quite difficult for people to withdraw their participation. Regardless of what researchers tell them about stopping at any time, saying "I want to stop now" is quite difficult to do. In FTF, if a participant does withdraw, one can ensure he or she is adequately debriefed. However, in much online research, dropping out is much easier: The participant just closes the browser window, and that is the end of it. As researchers, we may not even know how many people begin our experiments but do not complete them (although it is good practice to implement ways of discovering this). Those who drop out in this way will not receive debriefing, which again raises problematic issues. One potential solution is to use a "Quit" button on every page, which participants can use to terminate their involvement. Clicking on this would bring up a debriefing page. Although a number of authors have suggested using such mechanisms, there has been little evaluation of how well they work. Frick, Neuhaus, and Buchanan (2004) found that when a "Quit" button was provided, it was used by 57% of dropouts in their study.

6. CONTROL ISSUES: MINORS

Many online studies adopt open recruitment procedures whereby anyone coming across a recruitment notice may participate. This leads to a loss of control over who participants are: It is difficult to exclude people for whom the study might be considered inappropriate. Probably the largest class of people this concerns are children and minors, whom one should not expose to adult research materials that could be distressing or otherwise inappropriate. There are also issues around consent: Ethical codes usually require parental consent for minors to be involved in research. Identifying and contacting parents may be difficult, especially in an unrestricted recruitment procedure. Unless one uses strong identity verification procedures (e.g., asking for Social Security numbers, credit card details, or similar), one cannot guarantee minors will not take part. Even if one asks ages, and subsequently attempts to screen out minors or present age-appropriate alternative materials, there is no guarantee that respondents will answer truthfully. Probably, the safest course of action is to make sure that any materials that can be openly accessed are suitable for all.

There may also be instances in which minors are actually the intended participants in a project (e.g., in developmental or educational psychological research). In such cases, it is likely that special measures will need to be implemented. Reips (1999) presented case studies of such projects and ideas about ways to obtain parental consent. Other techniques worth examining include sampling from known populations (e.g., research panels composed of children) rather than open recruitment.

7. INDUCEMENTS AND REWARDS

The use of inducements and rewards has a long history in psychology, and in many cases their use has facilitated the conduct of research (see chap. 14, this volume). This is ethically acceptable as long as the inducement is not so great it would influence people to do something that they would normally prefer not to. A number of researchers have successfully used such rewards in online settings (for an example, see Baron & Siepmann, 2000). However, a number of issues must be considered. To pay someone—whether in cash, by entry into a lottery, or through some other kind of reward—one usually needs identifying information about them (e.g., name, address, Social Security number). Gathering this personal information creates additional issues (see the following section). Furthermore, one must consider whether mechanisms must be put in place to enable payment to people who drop out before the end of the experiment.

8. DATA SECURITY AND CONFIDENTIALITY: TECHNICAL AND LEGAL ISSUES

When personal information has been gathered, it is important to ensure that data are transmitted and stored in a secure way. Many online researchers use sites and transmission methods that are not secure and could allow interception of data (Reips, 2002). Researchers should endeavour to use more secure methods (for recommendations, see chap. 15, this volume, and Reips, 2002) and to minimize collection of personal information. This last point is important for another reason: In many countries, collection of personal information, its storage, and who has access to it are subject to legislation. Researchers should ensure they are familiar, and in compliance, with local legislation.

9. PUBLICATION OF QUOTATIONS AND PSEUDONYMS

When qualitative analyses are published, it is customary to provide quotations from participants (in interviews or in "found" text such as material posted on the Internet). It is also customary to disguise the identities of participants. In the age of powerful search engines, this becomes difficult. With much (if not most) text found online (e.g., a Usenet posting, a blog entry), it is possible to simply type the quoted text into a search engine and locate the original source and perhaps even the identity of the person who posted it. There are also strong arguments that the handles should be accorded the same status as real names in publication and, therefore, be anonymized. In cases in which screen names themselves are the object of study, this is clearly problematic.

One option researchers may use is to disguise identities and prevent searching for quotes by using altered or composite quotes from multiple sources (e.g., Kraut et al, 2004). This would then raise queries around the authenticity of the analysis—an obvious dilemma. Another option would be to seek permission for publication from the sources of the materials. The main issue with that option is that it may not be possible to contact the originators of materials—especially in the case of older material (for which e-mail addresses may have changed or the person dropped out of an online forum where their material was posted) or ephemeral materials (e.g., a screen name adopted in a chat room). There is also a tension with the notion of "authors" versus "subjects," and the issue of whether people should have their identities disguised or be credited for their creative work. Bruckman (2002) suggested different levels of disguise that may be appropriate, but there are clearly some thorny questions here without easy answers.

10. VIOLATION OF PARTICIPANTS' ENVIRONMENTS

People use different online spaces for different reasons, and they have different expectations of them. Whereas one person may use a chat room or forum as a form of social loafing, others may be going to them for social or emotional support. Individuals' attitudes to online spaces vary: Some become very attached to what they see as their virtual "home" and fiercely resent the intrusion of researchers looking for participants or data. Although there is considerable history of research being done successfully and nonoffensively online, there are also numerous documented examples of negative reactions (e.g., to posting participation requests in newsgroups or forums, or researchers entering virtual environments such as multiuser dungeons [MUDs]). In such cases, people have objected to becoming objects of study, feeling a safe space has been violated (e.g., Bruckman, 2002; Eysenbach & Till, 2001).

The distinction between private and personal space is important here, as is the perception of the inhabitants of those spaces. So how do researchers know whether their activities in a particular virtual space will be tolerated, welcomed, or rejected? Perhaps the best people to make that judgment are the inhabitants of the space themselves, and a number of authors have suggested that researchers should consider whether it is wise to seek permission from them and from owners or moderators of spaces if considered appropriate.

11. PROFESSIONAL RESPONSIBILITY TO COLLEAGUES

There is a requirement to make sure that only good research is conducted online. Because of its potentially high visibility, shoddily constructed or unethical research may be widely seen and damage the credibility of psychological research in general. For example, a number of authors have counseled against the use of deception online because it may perpetuate the impression that psychologists are tricksters, out to deceive, and that what they say cannot be taken at face value (Skitka & Sargis, 2006). Similar issues pertain to the recruitment of participants (especially from newsgroups or bulletin boards where participation requests are sometimes seen as spam or inappropriate) and failure to follow through on promises to disseminate study findings. Great care must be taken not to "spoil the pond" (Skitka & Sargis, 2006, p. 549).

12. FAILURE TO FOLLOW THROUGH ON PROMISES

Although it is important not to "spoil the pond" for colleagues by failing to follow through on promises to disseminate study findings, the most important responsibility in this respect is to participants. Often (if not usually) people participate in online research out of curiosity and are motivated by the promise of receiving feedback either on their own performance or on the study findings in general. To not provide this information is letting them down: "It is difficult to injure someone via the Web, except by dishonesty, so the fundamental ethical principle for Web-based research is honesty. If you promise some benefit for participation, then you must follow through and provide that benefit" (Birnbaum & Reips, 2005, p. 488). However, the nature of the feedback provided can also be an ethical concern (see Ethical Consideration 14).

13. DECEPTION

Codes of ethical conduct generally warn against the use of deception, outside of exceptional circumstances in which hypotheses cannot be tested in other ways and the scientific importance of the question is high. In the case of online research, there are additional reasons to counsel against the use of deception—issues of professional responsibility and the inability to debrief people who do not continue to the end of the study (APA Ethics Code Standards 8.07 and 8.08 require debriefing of participants after unavoidable deception). Researchers should consider whether deception is really required to achieve their research objectives, and if so, whether it is appropriate for the study to be conducted via the Internet at all.

14. GIVING FEEDBACK ON PERFORMANCE

Participants in Internet-mediated research often volunteer their involvement in return for some kind of feedback. For example, in much of Buchanan's work, participants completed online personality tests in return for being given feedback on their personality scores (e.g., Buchanan, Johnson, & Goldberg, 2005). However, one must be very careful about the type of feedback that is offered and given. One must also ensure that if participants are given information on their performance on a measure, it has been validated for use online and appropriate normative data are used to prevent possibly misleading feedback. It is possible that feedback from psychological tests can be distressing (e.g., being told one has a high Neuroticism score, a lower IQ than one expected, or some form of psychological problem). This is problematic, especially as researchers have limited ability to check that participants have understood feedback and are not adversely affected by it. In many cases, it may be wise to withhold detailed personalized feedback and provide only generic information about the study (having of course told participants that this will be the case during the informed consent procedure).

15. DEFINITIONS AND PERCEPTIONS OF PUBLIC AND PRIVATE SPACE

A number of authors (e.g., Kraut et al, 2004; Walther, 2002) have presented strong arguments that people who post messages in virtual spaces accessible to others (e.g., blogs, Web pages, Usenet) should be considered as behaving in public. Their comments are thus available for analyses in the same way as, say, personal ads placed in a newspaper would be. People posting material online are not only aware that their words are open to the public view, but they actually make the posting with the expectation that it will be viewed by an audience. However, there is evidence that some (perhaps many) people have different perceptions and expectations of online privacy, and perhaps of the audiences they want and expect. When planning research using materials people have posted online, one should consider what the poster's expectation of privacy was, whether or not that privacy actually exists.

16. EVIDENCE OF PARTICIPANTS' PSYCHOLOGICAL PROBLEMS

In some forms of research (e.g., online interviews dealing with sensitive topics), it is possible that researchers will encounter evidence of people who have psychological problems (e.g., people who are depressed or suicidal, or who have problems of which they themselves are unaware). This is especially the case given that levels of self-disclosure are often found to be high online. What does the researcher do in such a situation? Investigators should consider whether they are likely to be faced with such information and should develop contingency plans for such situations (e.g., will it be possible to refer the participant to an appropriate counselor or health care provider?).

Different Concerns for Different Techniques

Skitka and Sargis (2006) outlined three broad categories into which online studies may fall: *translational* (i.e., for which conventional procedures are adapted for the Internet), *phenomenological* (i.e., study of the medium itself), and *novel* (i.e., procedures that take advantage of the technical and social features afforded by the online world). Within these broad categories, one may identify more specific designs. For each of these different types of online research, different ethical considerations may be salient, and different measures may be adopted to deal with those ethical issues. In the material that follows, the key ethical issues for each type of research are indicated using the numbers from the preceding section. Of course, the extent to which each particular research project is subject to these concerns will vary, and specific projects may raise unique issues, so this list should be taken as indicative rather than comprehensive and prescriptive.

- *Surveys and other self-report questionnaires:* Includes cross-sectional and longitudinal panel designs; addressed in chapters 10 and 12 of this volume; Ethical Considerations 1, 2, 4, 5, 6, 7, 8, 11, 12, 14, and 16 are most likely to be relevant.
- *Experimental designs in which stimuli varying across conditions are presented and various reactions recorded:* Addressed in chapter 13; Ethical Considerations 1, 2, 4, 5, 6, 7, 8, 11, 12, 13, and 14 are most likely to be relevant.
- *Ability tests and measures of cognitive function:* Addressed in chapter 9; Ethical Considerations 1, 2, 4, 5, 6, 7, 8, 11, 12, 13, 14, and 16 are most likely to be relevant.
- *Interviews (with live or automated interviewers):* Ethical Considerations 1, 2, 4, 5, 6, 7, 8, 9, 11, 12, and 16 are most likely to be relevant.
- *Online focus groups:* Ethical Considerations 1, 2, 4, 5, 6, 7, 8, 9, 11, 12, and 16 are most likely to be relevant.
- *Study of Web browsing behavior (e.g., tracking use of sites, counting hits on Web sites):* Ethical Considerations 11 and 15 are most likely to be relevant.
- *Analyses of published or archived data sources (e.g., messages posted to bulletin boards, blogs):* Addressed in chapters 6 and 8. Ethical Considerations 2, 3, 8, 9, 10, 11, 15, and 16 are most likely to be relevant.
- *Live observation in interactive environments (MUDs, chat rooms), including participant observation:* Addressed in chapter 7; Ethical Considerations 1, 2, 3, 8, 9, 10, 11, 13 (if the researcher is covert), 15, and 16 are most likely to be relevant.
- *Analyses of online social networks:* Ethical Considerations 2, 9, 11, and 15 are most likely to be relevant.

Ethical Considerations 10 and 15 may apply to other situations, depending on recruitment method. For example, a recruitment message posted in a forum or Usenet newsgroup could be perceived as an invasion of privacy, regardless of whether the researcher is using a questionnaire or wants to recruit people for a focus group.

Ethical Scrutiny of Research

When deciding to run an online study, the researcher will need to consider all of the previously mentioned 16 ethical considerations. These issues will affect methodology and the strength of the research design. By taking these points into consideration, researchers will be able to defend the use of an online study to skeptics, draw more detailed conclusions regarding their results, and ensure the ethical use of this unique tool in conducting research. However, researchers are not the only ones who review study design and scrutinize it. Funding agencies and IRBs will both want additional information regarding how online studies provide increased strength to a design and how the study will be conducted in an ethical manner. Some of the areas that may be examined very specifically are described next.

In the United States, it is the duty of IRBs to review research and ensure the safety and well-being of participants. Hence, most IRBs may have particular concerns regarding online research, both out of lack of knowledge regarding how online research is conducted and because control and direct in-person access to the participants are often lacking.

Regarding informed consent, the Code of Federal Regulations in the United States (U.S. Department of Health and Human Services, 2001) states that informed consent may not be required if there is minimal risk to participants and if the study could not reasonably be carried out without waiving informed consent. The American Psychological Association (APA) Code of Ethics (APA, 2002; http://www.apa.org/ethics/code2002.html) states that informed consent may only be dispensed with if the study is anonymous and does not cause distress (Standard 8.05). Can online studies be designed to meet these requirements and do away with informed consent? Surely, they can; however, the burden is on the researchers to demonstrate this to their respective IRBs.

In cases in which informed consent is needed for online research, issues become more problematic. Some IRBs may accept an online Web-page–based informed consent that has an "Accept" button (taking the participant to the study) and a "Decline" button (taking them out of the study). However, some IRBs may require informed consent statements to be printed and mailed or faxed back to the researcher.

A related issue is ensuring that participants understand informed consent. In studies addressing particularly sensitive information, determining that participants understand what they are agreeing to may be deemed essential. There are several ways this can be addressed that should satisfy most IRBs. First, ensuring that the informed consent is clear and uses simple language, and avoiding the use of scientific or professional jargon as much as possible, will enhance understanding. Second, providing an actual test of understanding in the form of a few simple questions related to the IRB provides some evidence that participants read and understood the form. Finally, in nonanonymous studies, participants may be required to provide proof of age or identity. This final action, although likely not necessary for most online studies, could be needed in studies of psychotherapy outcome, for diagnostic verification, and studies addressing issues not suitable for minors.

Debriefing is often another area IRBs may require, although it may not be required by the Code of Federal Regulations. In addition, the APA Code of Ethics requires that debriefing information be available to all participants (Standard 8.08). How do researchers conducting online studies ensure access to debriefing? In online studies for which debriefing is required, how do researchers ensure that participants are adequately debriefed? For the first statement, whereby it is necessary to simply provide availability to debriefing information, the researcher can provide links to this information at the conclusion of the study and also provide this information if the participant decides to drop out of the study. In online studies, for which debriefing of all participants is required, researchers are left with fewer options. The only way to ensure adequate debriefing in this situation is to have contact information for the participant and for the researcher to contact the participant directly on conclusion of the study. This could be done in person, on the telephone, or through Internet-based media such as instant messaging or e-mail. However, these studies would by necessity not be anonymous studies.

Deception is problematic even in lab-based research. Using deception in online studies must be considered very carefully. Because of the difficulties in ensuring adequate debriefing, deceptive studies may be approved by IRBs only if the deception is minimal and if risk of distress to the participant is minimal. Even in this case, some IRBs may not approve online studies using deception.

One way in which online studies may actually provide an advantage over lab studies is in allowing participants to easily withdraw. To withdraw from or drop out of an online study, participants can simply close their Web browser or turn off their computer. In fact, drop-out rates are of concern with online studies and should be tracked to provide useful information to the researcher regarding their study. As mentioned earlier in the chapter, one way to accomplish this is to provide a "Quit" button that can track where and when the participant dropped out (Frick et al., 2004). But what if participants also want their data to be withdrawn from a study, even if the data are incomplete? This again can be accomplished through the use of a "Quit" button that can mark the data for exclusion or even deletion from the database. Otherwise, for nonanonymous studies, the participant can contact the researcher directly to request the removal of their information.

Obtaining participants can be problematic when conducting online research. The researcher can put up a study on the Internet and hope individuals will find it, but the response is likely to be low. Spamming e-mail lists and others is generally considered poor form and in some cases may actually be illegal. Inducements, such as cash awards, may attract participants, but, again, IRBs will want to ensure that coercion is not occurring and that the inducements are appropriate for the amount of time spent on the study. The topic of coercion is one that IRBs will pay particular attention to, and researchers who are also instructors or hold other positions of power must ensure that they do not implicitly coerce their students, clients, or patients.

Finally, another issue many IRBs may scrutinize is the security of the study data after and during collection. Although hacking and data security are mentioned often in popular media, for anonymous and nonsensitive information these are not areas of large concern. In these cases, even if the data are compromised by access of unauthorized individuals, the information is not identifiable with participants and is not likely to be of interest to anyone but the researchers. However, steps can be taken to reassure IRBs that the study data are at least as secure as data in a filing cabinet in an office. First, through the use of SSL (Secure Sockets Layer), the actual transmission of the data is encrypted and unreadable to anyone intercepting it on the Internet. Second, with the responses coded, the data in storage are unlikely to be understandable to others. Third, providing a data storage environment that is not directly connected to the Internet reduces the likelihood that others will be able to access it. Finally, for very sensitive information, data can be encrypted on storage and also removed daily from the server and stored in a secure physical environment.

These are just some of the areas with which IRBs and others may have concerns regarding online research. Whenever a new technology is used to collect data and solicit participants, it will be incumbent on the researcher to assure others that the research is being conducted ethically. And with the Internet, many tools are available. Researchers could use e-mail, instant messaging, blogs, e-mail lists, chat rooms, and others to both conduct research and solicit participants. As new technologies for the Internet emerge, new issues will arise.

Conclusion

In conclusion, we adopt the position that Internet-mediated research is ethically no more problematic than traditional ways of working (cf. Kraut et al, 2004). Indeed, one can point to ethical advantages arising from use of the Internet: For example, it is easier to ensure anonymity over the Internet than in FTF research; withdrawal from the research process is easier; and people may feel less worried about disclosing sensitive information (e.g., Turner et al., 1998).

However, different ethical considerations exist, and different measures may need to be taken to control for them in online and offline research. Not all types of Internet research are the same, and not all types of research are suitable for conducting online. Although there is emerging consensus about some of the key issues, there is still disagreement about some. There is also considerable scope for different ways of satisfying concerns for different projects—there can be no "one-size-fits-all" set of rules. In closing, our recommendation would be that researchers consider in turn each of the ethical considerations itemized earlier in this chapter, consider the extent to which each issue is applicable to their project, and implement measures to ensure their work meets the highest possible standards. The Internet has made researchers powerful, and we must be careful to use that power in a responsible way.

Additional Resources

Association of Internet Researchers. (2002). *Ethical guidelines for Internet research*. Retrieved March 11, 2005, from www.aoir.org/reports/ ethics.pdf.

> The Association of Internet Researchers' ethical guidelines adopt the position that ethical issues in online research can be ambiguous, perceived differently by different people. Rather than a set of strict guidelines, they incorporate a series of questions for researchers to ask themselves. Supplementary materials (including case studies and sample consent forms) are appended.

Ess, C. (2007). Internet research ethics. In A. N. Joinson, K. McKenna, T. Postmes, and U.-D. Reips (Eds.), *Oxford handbook of Internet psychology* (pp. 485–499). Oxford, England: Oxford University Press.

> This chapter provides a thoughtful discussion of research ethics, with consideration of underlying philosophical approaches. There is discussion of cross-disciplinary and cross-cultural issues. The chapter is more theoretical than most of the other documents cited but is still firmly rooted in practice with numerous real examples.

Frankel, M. S., & Siang, S. (1999, November). *Ethical and legal aspects of human subjects research on the Internet: A report of a workshop*. Retrieved, May 15 2002, from http://www.aaas.org/spp/dspp/sfrl/projects/intres/ report.pdf.

> One of the earliest attempts to provide a comprehensive overview of ethical issues in online research was a report of a workshop convened by the American Association for the Advancement of Science and the U.S. Office for Protection from Research Risks. It provides guidance for researchers and IRB members and has been influential. Alongside this document, one should probably read Walther (2002), who is very critical of the report, arguing that its recommendations overextend the authority of IRBs to instances that are not actually "human subjects" research. Walther raises concerns that this acts to stifle research.

Kraut, R. E., Olson, J., Banaji, M., Bruckman, A., Cohen, J. & Couper, M. (2004). Psychological research online: Report of Board of Scientific Affairs' Advisory Group on the conduct of research on the Internet. *American Psychologist, 59,* 105–117.

This article, the report of the APA's Board of Scientific Affairs' Advisory Group on the conduct of research on the Internet, presents an overview of relevant issues and places them in the context of the U.S. IRB system. A key document for online researchers, it presents clear advice to researchers and IRBs and should be considered required reading, even for those working outside the United States.

O'Dell, L., Bartram, D., Buchanan, T., Hagger-Johnson, G., Hewson, C., Joinson, A. N., Mackintosh, B. (2007). *Report of the Working Party on Conducting Research on the Internet: Guidelines for ethical practice in psychological research online.* Retrieved November 21, 2008, from http://www.bps.org.uk/downloadfile.cfm?file_uuid=2b3429b3-1143-dfd0-7e5a-4be3fdd763cc&ext=pdf.

The British Psychological Society guidelines for online research identify a number of key considerations that may present issues for online research projects (verifying identity, understanding of public and private space, informed consent, levels of control, withdrawal, debriefing, deception, monitoring the consequences of research, protecting participants and researchers, and data protection). They present a typology of different research designs in which participants are either recruited or unaware that they are being observed, and are either identifiable or anonymous. They argued that different considerations will apply to different conditions, and they made suggestions as to how researchers may overcome these ethical challenges.

References

American Psychological Association. (2002). Ethical principles of psychologists and code of conduct. *American Psychologist, 57,* 1060–1073.

Association of Internet Researchers. (2002). *Ethical guidelines for Internet research.* Retrieved March 11, 2005, from http://www.aoir.org/reports/ethics.pdf

Baron, J., & Siepmann, M. (2000). Using Web questionnaires for judgment and decision making research. In M. H. Birnbaum (Ed.), *Psychological experiments on the Internet* (pp. 235–265). New York: Academic Press.

Birnbaum, M. H., & Reips, U.-D. (2005). Behavioral research and data collection via the Internet. In R. W. Proctor & K.-P. L. Vu (Eds.), *Handbook of human factors in Web design* (pp. 471–491). Mahwah NJ: Erlbaum.

Bruckman, A. (2002). Studying the amateur artist: A perspective on disguising data collected in human subjects research on the Internet. *Ethics and Information Technology, 4,* 217–231.

Buchanan, T., Johnson, J. A., & Goldberg, L. (2005). Implementing a five-factor personality inventory for use on the Internet. *European Journal of Psychological Assessment, 21,* 115–127.

Eysenbach, G., & Till, J. E. (2001). Ethical issues in qualitative research on Internet communities. *British Medical Journal, 323,* 103–105.

Frick, A., Neuhaus, C., & Buchanan, T. (2004, March). *Quitting online studies: Effects of design elements and personality on dropout and nonresponse.* Poster session presented at German Online Research '04, Duisburg, Germany.

Kraut, R. E., Olson, J., Banaji, M., Bruckman, A., Cohen, J., & Couper, M. (2004). Psychological research online: Report of Board of Scientific Affairs' Advisory Group on the conduct of research on the Internet. *American Psychologist, 59,* 105–117.

O'Dell, L., Bartram, D., Buchanan, T., Hagger-Johnson, G., Hewson, C., Joinson, A. N., Mackintosh, B. (2007). *Report of the Working Party on Conducting Research on the Internet: Guidelines for ethical practice in psychological research online.* Retrieved November 21, 2008, from http://www.bps.org.uk/downloadfile.cfm?file_uuid=2b3429b3-1143-dfd0-7e5a-4be3fdd763cc&ext=pdf

Reips, U.-D. (1999). Online research with children. In U.-D. Reips, B. Batinic, W. Bandilla, M. Bosnjak, L. Gräf, K. Moser, & A. Werner (Eds.), *Aktuelle Online-Forschung—Trends, Techniken, Ergebnisse* [Current Internet science—Trends, techniques, results]. Zürich: Online Press. Retrieved December 2, 2008, from http://gor.de/gor99/tband99/pdfs/q_z/reips.pdf

Reips, U.-D. (2002). Internet-based psychological experimenting: Five *do*s and five *don't*s. *Social Science Computer Review, 20,* 241–249.

Skitka, L. J., & Sargis, E. G. (2006). The Internet as psychological laboratory. *Annual Review of Psychology, 57,* 529–555.

Turner, C. F., Ku, L., Rogers, S. M., Lindberg, L. D., Pleck, J. H., & Sonenstein, F. L. (1998, May 8). Adolescent sexual behavior, drug use, and violence: Increased reporting with computer survey technology. *Science, 280,* 867–873.

U.S. Department of Health and Human Services. (2001). *Code of Federal Regulations: Protection of human subjects.* Washington, DC: Author. Retrieved September 15, 2004, from http://www.hhs.gov/ohrp/humansubjects/guidance/45cfr46.htm

Walther, J. B. (2002). Research ethics in Internet-enabled research: Human subjects issues and methodological myopia. *Ethics and Information Technology, 4,* 205–216.

Index

About the Editors

Samuel D. Gosling, PhD, is a professor of psychology at the University of Texas at Austin. He has been using the Internet to collect data since the mid 1990s, when he created a questionnaire to collect personality ratings of pets by their owners. Since then, he has published numerous articles that make use of data collected on the Internet; these articles focus on such diverse topics as personality change over the life span, the links between music preferences and personality, geographic variation in psychological traits, and perceptions of others based on their Web sites and their online social networking profiles (e.g., Facebook). His 2004 *American Psychologist* article focused on evaluating the pros and cons of Internet methods. Dr. Gosling's substantive research has focused on animal personality and on how human personality is manifested in everyday contexts like bedrooms, offices, clothing, Web pages, and music preferences. The latter topic was summarized in his book, *Snoop: What Your Stuff Says About You* (2008). Dr. Gosling is the recipient of an American Psychological Association's Distinguished Scientific Award for Early Career Contribution.

John A. Johnson, PhD, is a professor of psychology at the Pennsylvania State University, DuBois. He also serves as the consultant for the International Personality Item Pool, a Web-based repository for psychological measures in the public domain. He entered the field of computer-assisted psychological research in 1986, when he wrote microcomputer programs for scoring and interpreting the Hogan Personality Inventory. When the World Wide Web emerged in the 1990s, he transported concepts from these programs to the Web. He has published research on assessing the validity of data collected on the Internet and on sharing data through Web-based collaboratories.